"Jehane Ragai combines scientific expertise with aesthetic sensitivity. In consequence, the first edition of *The Scientist and the Forger* offered an insightful new perspective on an intrinsically murky topic. This second edition is rendered even more fascinating by the inclusion of case studies that illuminate the complexities, tensions and ambiguities of the fast-expanding international art market. Even those coming completely fresh to the subject will be fascinated by this engrossing and finely written book."

Martin Rees (Lord Rees of Ludlow) OM, FRS, FREng, FMedSc
Astronomer Royal, Former President of the
Royal Society and Former
Master of Trinity College, Cambridge

"Seen with the eye of the scientist but addressed to a wider public, Jehane Ragai's *The Scientist and the Forger; Insights into the Scientific Detection of Forgery in Paintings* (2015) covers the whole spectrum of scientific authentication methods, illustrating them with a wide range of examples. The new revised edition, with the subtitle *Probing a Turbulent Art World*, includes the most recent cases and devotes new attention to all those involved, thus presenting a holistic view of the art world, the psychology of the forger and the role of scientific investigations in the debates. Even the seasoned art historian will learn much from this informative book."

Jean-Michel Massing
Professor of History of Art, King's College, Cambridge

"I enjoyed very much reading Jehane Ragai's first book *The Scientist and the Forger* which, in a scholarly but at the same time most entertaining, seductive way, brought me into the world of art forgery (new to me) and modern forensic technology (where I am as a physical chemist more at home) with numerous exciting details about how each mystery was solved and the culprit eventually brought to justice. Professor Ragai's second book — which I was very eager and curious to read as I anticipated that interesting things may have happened — indeed was not a disappointment, as in her hands the topic has exploded like an expanding universe. There is a richness in detail from a selection of cases, and the author is a good story teller. As those who are more initiated in the field no doubt will notice there are noteworthy new developments that have taken place in relation to forgery and authentication cases since the first edition. The psychological dimension is particularly interesting and her own original interpretation of the psychological behaviour of Ann Freedman, the gallery president in the Knoelder Gallery scandal, will probably win general acceptance. I can warmly recommend the second edition of *The Scientist and the Forger* which is a great book."

Bengt Nordén
Professor, Chemistry and Chemical Engineering, Physical Chemistry, Member Royal Society of Arts and Sciences in Gothenberg and Member of the Royal Swedish Academy of Sciences

"Forgery in paintings is a large and increasing problem. It is probably only when the judgement of art-historians is combined with objective scientific analysis of a painting that we can be certain if it is genuine or a forgery. This beautifully written and illustrated book covers both the scientific techniques used to detect forgeries and the plight of collectors, gallery dealers and authenticators. It gives helpful case studies of the detection of forgeries, many of them recently discovered. This clearly written and informative book should be on the bookshelves of every art lover, painter, collector, art expert, gallery dealer, auction house, art lawyer and authenticator. If you can purchase only one book on art forgery, buy this one!"

Sir Colin Humphreys CBE, FREng, FRS
Goldsmiths' Professor of Materials Science
University of Cambridge

"As an artist, forgery was my natural interest in this book, but Jehane Ragai incisively opened all the fascinating, unexpected worlds behind her subject — science, sleuthing and skulduggery. Surely there is no more thorough and entertaining scrutiny of the art of forgery."

June Mendoza AO, OBE, RP, ROI, HonSWA
Member of the Royal Society of Portrait Artists

"The book by Professor Jehane Ragai is a wonderful example of the unity of culture with no sense in claiming Snow's dualism between the two cultures: sciences and humanities. The writer deals with the issue of art, forgery and the fundamental role of science and technology with excellent sagacity allowing many readers — not only skilled or experienced people — to understand the relationship between science and art in the complex world of forgery. The author, a professor of chemistry, succeeds thanks to her background rich in the knowledge of matter and its infinite transformations, to also appreciate the material feature associated with works of art. Indeed, *Mother-Matter* — as Primo Levi called it — is at the basis of everything, and of course also of art. Reading this book we appreciate once more this truth!"

Luigi Dei
Professor of Environmental and Cultural Heritage Chemistry
Rector of the University of Florence, Italy

"Through a fascinating series of case studies of artists from the 15th to the 20th century, Ragai shows how art historians and scientists can collaborate conclusively to authenticate paintings or demonstrate that they are forgeries. As prices for works of art soar, we learn how problematic forgery has become for commercial galleries and auction houses, and of the downfall of a venerable gallery, Knoedler, whose president threw caution to the winds. Ragai concludes with a solution for new works of art: biologically engineered DNA finger-printing! She is an excellent story-teller with an interest in the psychology of forger and accomplice which makes this important book a pleasure to read. It fills a void in the literature and should be in every art-historical library."

Vivien Perutz
Visiting Lecturer, Department of History of Art
University of Cambridge

"Jehane Ragai balances deftly principles of *caveat emptor* ('buyer beware') with the professional negligence implications to art experts validating the provenance of art. Her forensic review of case law, and intimate understanding of the science, reveals a judicial reluctance to set a high threshold before a breach of a duty of care owed by an art expert to a prospective buyer is established. Courts are reluctant to impose a duty on experts to apply a full array of often expensive scientific techniques to validate the provenance of art, ahead of sale. The message to take away from Jehane's lively prose is that buyers of expensive works of art need to read this book to understand the science underpinning the detection of art forgery before contemplating buying expensive art."

Marcus Grant
Specialist Insurance Fraud Barrister

The Scientist and the Forger

Probing a Turbulent Art World

Second Edition

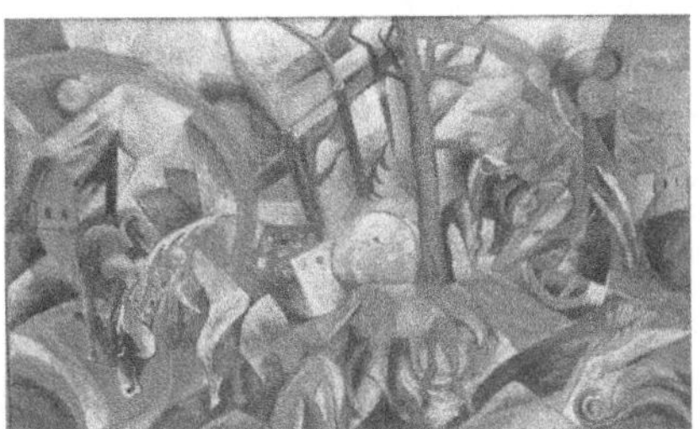

The Scientist and the Forger

Probing a Turbulent Art World

Second Edition

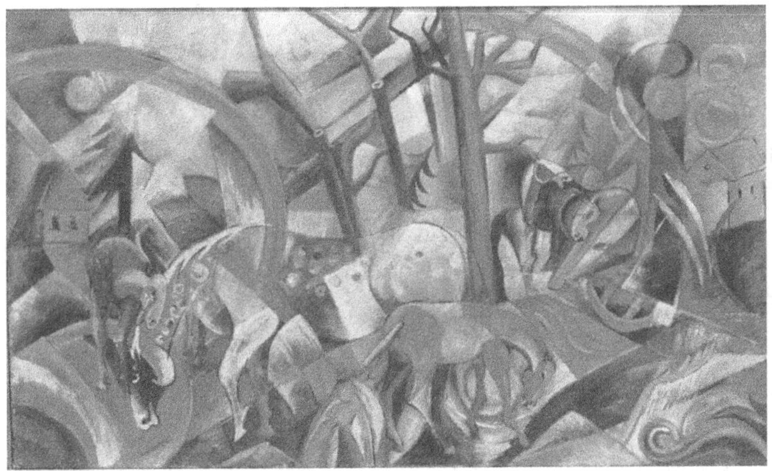

Jehane Ragai

The American University in Cairo, Egypt

World Scientific

NEW JERSEY • LONDON • SINGAPORE • BEIJING • SHANGHAI • HONG KONG • TAIPEI • CHENNAI • TOKYO

Published by

World Scientific Publishing Europe Ltd.
57 Shelton Street, Covent Garden, London WC2H 9HE
Head office: 5 Toh Tuck Link, Singapore 596224
USA office: 27 Warren Street, Suite 401-402, Hackensack, NJ 07601

Library of Congress Cataloging-in-Publication Data
Names: Ragai, Jehane, author.
Title: The scientist and the forger : probing a turbulent art world / by
Jehane Ragai (The American University in Cairo, Egypt).
Description: 2nd edition. | New Jersey : World Scientific, 2017. | Includes index.
Identifiers: LCCN 2017030466 | ISBN 9781786344205 (hc : alk. paper)
Subjects: LCSH: Painting--Forgeries. | Art--Psychology.
Classification: LCC ND1660 .R34 2017 | DDC 751.5/8--dc23
LC record available at https://lccn.loc.gov/2017030466

British Library Cataloguing-in-Publication Data
A catalogue record for this book is available from the British Library.

First published 2018 (Hardcover)
Reprinted 2018 (in paperback edition)
ISBN 978-1-78634-594-3 (pbk)

For any available supplementary material, please visit
http://www.worldscientific.com/worldscibooks/10.1142/Q0123#t=suppl

Desk Editors: Anthony Alexander/Jennifer Brough/Shi Ying Koe

Typeset by Stallion Press
Email: enquiries@stallionpress.com

For my husband John
and daughters
Nazli and Heddy
with love.

Contents

About the Author

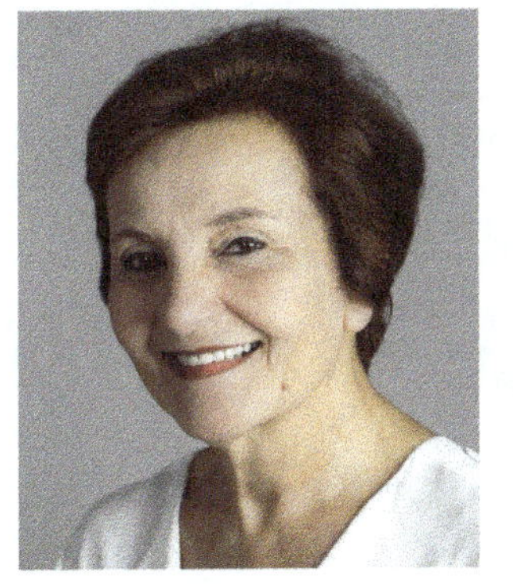

Dr Jehane Ragai, a surface scientist who studied for her PhD in London, is an emeritus Professor of Chemistry at the American University in Cairo.

She has lectured extensively in the US, Europe and the Middle East to university and museum audiences on the scientific detection of forgery in paintings and on topics related to Ancient Egyptian Science, in particular on colour.

Owing to her additional interest in archaeological chemistry, she became a consultant to the American Research Center in Egypt (ARCE) Sphinx project. She has served on the National Committee for the Study of the Sphinx, and from 2001–2008 was a member of the Board of Governors of the ARCE.

Since 2008 Dr Ragai has been a jury member for the L'Oréal-UNESCO Women in Science Award, founded by the Nobel laureates Christian de Duve and Pierre-Gilles de Gennes.

As a faculty member in the Department of Chemistry of the American University in Cairo, she has chaired its Senate, its Department of Chemistry and was the Director of its Chemistry Graduate program. The recipient of several AUC Trustees merit awards, Dr Ragai also received the School of Sciences and Engineering award for her role as chair of the Department of Chemistry and in 2013 she was awarded the university-wide best teacher award. She was recently elected as a foreign member of the Royal Swedish Society of Arts and Sciences in Gothenburg (a multidisciplinary research academy established in 1778).

She lives mostly in Cambridge, UK (with her husband) and partly in Cairo.

Foreword to the Second Edition

This second edition of Jehane Ragai's *The Scientist and the Forger* (first edition 2015) is a welcome addition to the field of technical research into the authenticity and condition of paintings. The chance of (re) discovering an original or unmasking a fake is essentially a battle between the historical creators and modern professional researchers who pursue objective analysis. Jehane Ragai's publication increases knowledge on measuring techniques, and builds bridges between the different fields of art history, restoration sciences and materials science.

When confronted with a work of art, its beauty and mystery can appeal to a euphoric state of mind and create an undefinable impact on the individual. Its authenticity and condition form its solid foundation. Despite the different interests and end goals of professionals studying, trading and preserving works of art, all of them contain a singular common ground: authenticity and condition.

Authenticity and condition are embedded in how a piece of art is created. The materials speak to the history of its origin and existence, and are the so-called primary source. Secondary sources, including proofs of ownership or authentication statements, can provide an indication, but do not adequately prove the moment of creation and can only give partial information of the work's journey through time. Whether or not a work of art is authentic, a copy or intended as a forgery, it carries along a past that is discoverable in its physical, chemical and technical properties. These properties apply to the work as a whole, as well as to all partial elements. Whether it is with good or bad intentions, a copyist or forger trying to influence the course of history cannot recreate the moment in time of the actual creation of the work. However good the technique or knowledge of a later creator may be, he or she *knows* that there will always be an omission: an unchangeable or inalienable feature that is only to be found in the original. It is this feature or combination of features that is found in the chemical, physical and technical properties of the material. Depending on the experience and expertise of a forger, the materials used for an object can be put into the correct timeline and historical context.

Materials used for works of art include a "materials provenance", which includes its composition, source date of production and distribution. It almost goes without saying that the collecting and interpreting of "materials provenance" should be done based on an insightful protocol in order to be reliable and viable for comparative research. The reference material, too, should be unambiguously selected and analysed. In order to be effective and academically objective the research protocol, for each time period and/or artist, should be clearly described in transparent discourse so contemporary reviews are reliable, and future insights and findings can permanently improve the historical context, its current condition and its economic value.

To achieve optimal results, authenticity research is ideally carried out within a triangle of disciplines: art history, restoration and materials science. These seemingly contrasting disciplines are not each other's counterparts, but instead reinforce each other in the search for historical knowledge, the determining of economic value and the improvement of future conservation. Making use of all three disciplines' different qualities optimizes the understanding for both the individual collector and the art world as a whole.

In the future, the use of modern analysis techniques for determining authentication and condition is essential. The naked eye alone is no longer sufficient for critical analysis. The value of art objects has grown exponentially, and counterfeiters and forgers are more talented and resourceful than ever. Using advanced techniques, they seek out the weakest link and act accordingly to create realistic and convincing forgeries or copies. Cooperation between the art historian, the restorer and the materials scientist is therefore essential to provide irrefutable evidence needed to classify the work under investigation. This starts in academic training of the three professions and moves into the commercial market once these individuals are qualified — it is essential therefore that each are made aware of the important role of the other within the global art (black)market, and the skills and expertise each brings to the table in case of the right classification of a forgery, copy or the rediscovery of an original.

In this book, Jehane Ragai tackles these difficult and complex issues using case studies, scientific evidence and legal frameworks to demonstrate how the biggest forgeries of our time have helped develop techniques for detection.

Milko den Leeuw
Painting Restorer & Researcher ARRS, The Hague
Congress Organiser, Authentication in Art Foundation ©
www.authenticationinart.org

Preface

In May 2016, I attended *The Authentication in Art Foundation Congress* held in The Hague. The congress, as described by its organisers, had "the ambition to set targets and to deliver guidelines and solutions to improve the field of authentication in art" with a special focus on paintings. A wide spectrum of international professionals involved in the art world — lawyers, insurers, art historians, scientists, educators and other key players — jointly addressed the most pressing dilemmas and proposed innovative solutions.

Their compelling presentations covered topics such as the lack of transparency in the art market and means of countering it, legal and art-historical perspectives with respect to Art Authentication, the plight of experts today, E-related initiatives that would address the problems of provenance and authenticity, Technical Art History and other forgery-related matters.

What attracted my attention was that scientific lore and forgery were the common themes directly or indirectly permeating all the subjects under discussion. Less than a year earlier, I had published the first edition of *The Scientist and the Forger* and realised that the issues surrounding forgery in paintings had moved so rapidly that an updated and somewhat different version of the book was called for.

The first edition had explored scientific techniques employed in the detection of forgery, but now there is an exigent need for a new rendition which would devote more attention to the plight of authenticators, collectors and gallery dealers, and to the relentless efforts made by professionals belonging to diverse disciplines to safeguard today's art market for future generations. Accordingly, I have striven to share noteworthy developments that have taken place in relation to forgery and authentication cases since the first edition.

At the Congress, I listened to discussions on the Knoedler scandal (page 84), and on La Bella Principessa (page 105), I wish to share with the reader my own interpretation of the psychology of the behaviour of Ann Freedman, the president of the Knoedler Gallery, and to suggest possible means of solving once and for all the long-standing enigma of La Bella Principessa.

It is in this spirit that I have endeavoured to present here, as succinctly as possible, a holistic approach to the multifaceted components of an art world shaped by an ever-changing chain of events and where forgery and science play a fundamental role.

Acknowledgements

I would like to record my indebtedness to the many people who helped me during the preparation of the first edition of *The Scientist and the Forger*, and to refer back with gratitude to each and every one of them mentioned in that publication.

With regard to the present second edition, I am indebted to all those who provided encouragement and support, offered comments and suggestions and granted me the permission to use their images.

First and foremost, I am extremely grateful to my daughter and exceptional copy-editor Nazli, who patiently went through the manuscript line by line, making numerous valuable comments and bringing coherence and clarity to the text.

Very special thanks go to the editorial board of World Scientific Publishing, and in particular to Laurent Chaminade for his continual encouragement and belief in my work, to Mary Simpson, Jennifer Brough and, in particular, Anthony Alexander for their invaluable help and assiduity in overseeing the final editing of this book, to Nigel Thompson and Jessica Clot for their great help in marketing and finally Shi Ying Koe for her patience in the designing of the cover.

I would like also to express my deepest gratitude to a list of scholars who have read sections (and often entire chapters) of this book and generously given me the benefit of their respective expertise: Marcus Grant (Temple Garden Chambers), Martin Kemp (University of Oxford), Dean Nicyper (New York City Bar Association), Robert Norton (Verisart), Vivien Perutz (University of Cambridge) and John Meurig Thomas (University of Cambridge).

I owe a special debt to Libby Sheldon (University College London) for clarifying the scientific techniques used in the Lucian Freud case and to Milko den Leeuw (Authentication in Art Foundation) and Laurence Schindell (ARIS Title Insurance) for their wonderful support.

I owe a special debt to my daughter Heddy for her invaluable help in identifying the right color for the cover of the book, to Fadia Badrawi for her patience and for her excellent drawings and to Gehan Ghali for her great assistance with the references.

I also wish to thank Art Roster, Mt Shasta, and Pascal Cotte (Lumiere Technology), for their generosity in allowing me to use their images pro bono in my book.

Producing this book has indeed been a joint effort, and I am grateful to the many people who have made it come to fruition; doubtless some mistakes and omissions may still be present. If indeed there are, the responsibility is entirely mine.

Last, but not least, I would like to pay tribute with gratitude and love to my dearest husband John, to my very special daughters Nazli and Heddy and to my wonderful sister Aziza. Thank you so much for the constancy of your support.

Acronyms

AMS	—	Accelerator mass spectrometry
CCD	—	Charge-coupled device
CNRS	—	French National Research Center
CT	—	Computer tomography
DNA	—	Deoxyribonucleic acid
DR	—	Digital radiography
EDX/EDXRF	—	Energy dispersive X-ray fluorescence
ESRF	—	European synchrotron radiation facility
ETHZ	—	Swiss Federal Institute of Technology, Zurich
FTIR	—	Fourier transform infrared spectroscopy
GC	—	Gas chromatography
HPLC-MS	—	High performance liquid chromatography-mass spectrometry
IFAR	—	International Foundation for Art Research
INFN	—	Instituto Nazionale di Fisica Nucleare
INSP	—	Paris Institute of Nanosciences
IRR	—	Infrared reflectography
IR	—	Infrared
MS	—	Mass spectrometry
L.A.M.	—	Technique (layer amplification method)
LA-ICP-MS	—	Laser-ablation-inductively coupled plasma–mass spectrometry
LBP	—	La Bella Principessa
MC-ICP-MS	—	Multiple collector-inductively coupled plasma–mass spectrometry

Micro-FTIR	—	Micro-Fourier transform infrared spectroscopy
MLF	—	Mona Lisa Foundation
OM	—	Optical microscopy
PIXE	—	Proton induced X-ray emission
PLM	—	Polarised light microscope
Py-GC-MS	—	Pyrolysis–gas chromatography–mass spectrometry
RRP	—	Rembrandt Research Project
SEM	—	Scanning electron microscopy
UCL	—	University College London
UV	—	Ultraviolet
UVF	—	Ultraviolet fluorescence
XCT	—	X-ray computer tomography
XRD	—	X-ray diffraction
XRF	—	X-ray fluorescence
XRR	—	X-ray radiography

Introduction

When money gets turned into art that is one thing, but when art gets translated into money, it's a problematic transaction.

Edward Helmore[1]

While problems of authenticity in paintings have existed for hundreds of years, over the last few decades they have reached untenable proportions. The scandalous scale of recently discovered forgeries together with a surprising rise in the price of artworks and a simultaneous increase in the knowledge and skills of the professional forger underscore the crucial role played by modern science in exposing artistic fraud.

The importance of sophisticated technical methods in unravelling fakes and in assisting in the authentication process has been demonstrated in the first edition of this book. As forgers become more knowledgeable and skilled, there is an ever-growing need for more advanced scientific approaches to scrutinize techniques of painting, track down manufacturers and identify new pigments. Meanwhile, contemporary artists intensify their quest to safeguard the authenticity of their works for present and future generations of collectors.

Today, a forger who creates a painting in the style of a known master and claims its authenticity is aware that the hoax may be uncovered as a result of the detection of features such as questionable underdrawings, the discovery of anachronisms in the pigments or even by something as simple as the scrutiny of surface cracks. Skilled forgers proceed with astounding ingenuity to avoid pitfalls such as these, and in doing so constantly present science with new challenges.

In any case, scientific scrutiny on its own cannot address all the problems associated with an art world plagued by a lack of transparency and teeming with unscrupulous so-called forensic scientists. It is a world which is faced with a complex situation arising from collectors looking for good investments in art, museums anxious to make the right assessment of an artwork but reluctant to resort to expensive investigations, dealers and agents keen to live by the sales in the art market, and claims of art fraud being met with legal libel. It is also relevant to recall that once tracked down, forgers are given astonishingly light sentences followed by celebrity upon release.

The present book attempts to equip the reader with a holistic understanding of an art world shaped both by history and by fast-evolving views and trends, one which is increasingly permeated by scientific lore and pervaded by the diverging standpoints offered by lawyers, insurers and other key players. In pursuit of an understanding of forgery and questions related to authentication, the reader is taken on a journey through a world where art and law have become inextricably linked, litigations are omnipresent, financial implications stir the highbrow artworld, attempts are made to safeguard a potentially dying art market, and education and information technology are called on to play a crucial role.

In this book, I have attempted to afford the reader a first glimpse into exciting stories of forgery detection and authentication. A window is provided onto the recently uncovered old Masters forgery scandal that rocked the art world, the machinations of former hippie artist/forger Wolfgang Beltracchi, the unparalleled rise in Russian avant-garde forgeries following Russia's turbulent 20th century with the unexpectedly dramatic case of a fake Chagall painting ending in a French furnace, the discovery by the National Gallery of hitherto unknown forgeries in its own collection, the thirty-two works allegedly painted by Pollock (2005) and discovered by Alex Matter the son of Pollock's good friend Herbert Matter in his parents' attic and onto other forgeries brought to light by scientific analysis.

In an art world afflicted by so many complexities, authentication facilitated by science brings welcome relief. A self-portrait long thought to have been painted by one of Rembrandt's former students was declared genuine, and when connoisseurs were reluctant to authenticate both of van Gogh's paintings *Sunset at Montmajour* and *Still Life with Meadow Flowers and Roses* on the grounds of style and aesthetic value alone, it was the successful convergence of art-historical evidence with scientific investigation that led to the final verdict of authenticity. Compelling scientific evidence supported the identification of a painting depicting *Saint Praxedis* as a Vermeer, and most unexpectedly, *Man in a Black Cravat* was attributed to Lucian Freud, despite this artist's adamant refusal to recognize it as one of his own. These and other cases of authentication demonstrate that art historical or connoisseurship working hand in hand with scientific investigation can bring about very good news indeed.

The second part of the book focuses on court cases revealing the plight of art experts, auction houses and collectors, with an account of some of the present-day efforts aiming to address such a predicament. A proliferation of lawsuits against authentication experts, gallery dealers, authors of catalogues raisonnés, auction houses and foundations providing candid opinions on artworks threaten the natural functioning of the art market to such an extent that unhindered, they could lead to its freezing and eventual downfall.

The predicament of *experts* is illustrated by the case of Elena Basner who, after giving a totally informal opinion in her home on a gouache attributed to Boris Grigoriev, was faced with an avalanche of events that culminated in a court case and in her arrest. Then there is the famous case of Caravaggio in which the owner, convinced by the evaluation made by the auction house agreed to sell it; when later, after the sale, a different opinion was made on the work, the seller pursued legal action. This case typifies the problems faced by the *auction house*. The plight of the *collector* is exemplified in the famous Knoelder trial in which a work by a renowned artist bought in good faith and without any further probing was later found to be a forgery. Most importantly, with regard to whether or not Ann Freedman, president of the Knoedler gallery, knowingly sold forged paintings, new light is shed on the psychology of her behaviour.

So, what can be done, asks the book, to reassure authenticators and auction houses that they will be protected from expensive, tortuous lawsuits? To remedy the situation, the New York City Bar Association has passed a bill on June 15, 2015, to protect experts. The bill has still to be approved by the New York Assembly and eventually by the New York Governor, but it remains to be seen if, based on this Act, authenticators will feel less vulnerable and sufficiently backed-up to resume their assessment activities. An answer to all these uncertainties may be provided by a recently developed DNA technique which would safeguard the art market and protect authenticators, a Holy Grail for all artists and collectors; or maybe, in the very near future, blockchain technology (which is fully explained in Part II) will revolutionise the art world by providing a far more secure system which will enable users to verify a work's authenticity, condition and provenance.

The dynamic and unforeseeable nature of the art world is reflected in a third part of the book that deals with recent discoveries related to Leonardo's *Mona Lisa* and two other works: The Isleworth and Prado Mona Lisas. What are the full implications of the recent revelation, in the underlayers of the Louvre *Mona Lisa,* of a portrait depicting a woman totally different from the Mona Lisa we know? Was the Isleworth *Mona Lisa* also painted by Leonardo or did a gifted artist try hard to emulate the master's style and add discreet variations, and is there real evidence for the 'two Lisas hypothesis'? Why is the Prado *Mona Lisa* a copy of particular interest?

This part also entails a discussion of the controversial *La Bella Principessa,* a portrait in ink and chalk on parchment, attributed by some to Leonardo da Vinci and suggesting a possible solution to this enigma. The claim by forger Shaun Greenhalgh's that he faked *La Bella Principessa* is analysed together with an evaluation of his psychological make up.

In a last part, the reader is introduced, in language easily accessible to the layperson, to a wide spectrum of old and new scientific methods for the study of paintings, progressing from the simple to the complex and entailing the examination of the surface, frame, body and paint layers of the artwork.

Scientific tests in isolation should not be the sole determinant of fraud, nor can connoisseurship alone play a definitive role in the evaluation of works of art. Intuition and a deep understanding of the work at hand, together with a close analysis of technique, appearance, style and design of the artwork are essential in complementing the objectively collected scientific data.

In the present context where we see a booming art market afflicted by a proliferation of forgeries, possible responses to the detection of forgeries in general and those of establishing the authentication of works by old masters in particular, provide more than a glimmer of hope in an art world afflicted by a steep rise in litigations and condemned by its manifest secrecy.

Part I

Attribution

Establishing the First Link in the Art Chain: Attribution

Being able to access numerous aspects by a scientific approach increases the fascination beyond the borders of pure art studies and brings together two fields of human activity that inspires and deepens the understanding of both fields: science and art.

Richard R. Ernst[1]

The Curator and Gallery Dealer

The curator and gallery dealer play a crucial role in the initial evaluation of a painting allegedly executed by a renowned artist. It is their responsibility to study the work's provenance by determining a sequence of ownership all the way back to the artist. Correct verification of each one of the links on the ownership chain can prove elusive however, especially in light of the convincing information forgers have been known to present on the work's origin, successive whereabouts and on the reasons for its obscurity up till then.

As pointed out by art historians Chartier and Notehelfer:

Provenance is probably the most critical element in authentication. There is no substitute for an iron-clad provenance back to the hand of the artist touching the canvas. However, this is relatively rare even in the most established of collections.[2]

Forgers generally invent a story of a destitute family that has owned the painting on offer for many generations and now needs to sell it while wishing to remain anonymous.[3] A price that is too good to be true might arouse the curator's suspicion.

The ingenuity and imagination of forgers have frequently shone as vividly through their plausible stories as they have in their fake masterpieces. If, in addition, the dealer presenting the work has seemingly impeccable credentials, even the most prominent experts can be fooled.

As early as 1925, German art dealer Otto Wacker[4] invented an alluring tale to prove that the van Gogh paintings he was selling were genuine. He claimed that they had been acquired by a Russian dealer from the sovereign dynasty who had transferred them to Switzerland and fearing reprisals from his relatives who were still living in Russia, wished to remain anonymous. Wacker who had developed a sound reputation in the art field easily convinced experts of his story and had them issue certificates of authenticity without any real proof of provenance.

Another very imaginative and daring approach was that of Elmyr de Hory (Fig. 1) who, in the 1960s, created paintings resembling the works of a spectrum of great masters (Picasso, Modigliani, Matisse, etc.). He removed the original folio photographs of their works from old art books and replaced them with photographs of his own paintings. Many potential buyers were taken in by seeing a photograph of the same painting they were being presented with in an ostensibly old art book.[5]

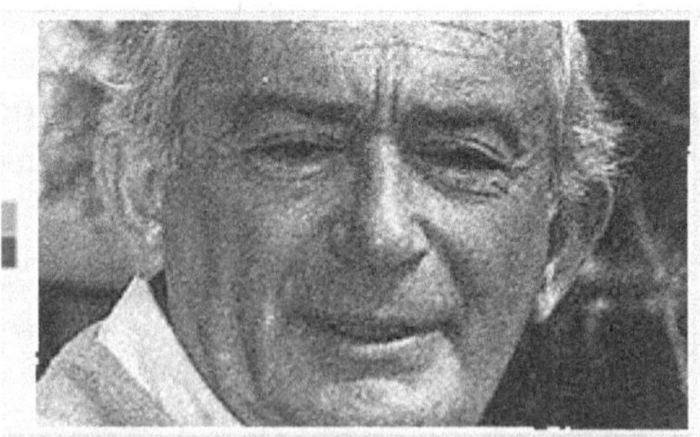

Fig. 1 Elmyr de Hory (1906–1976)

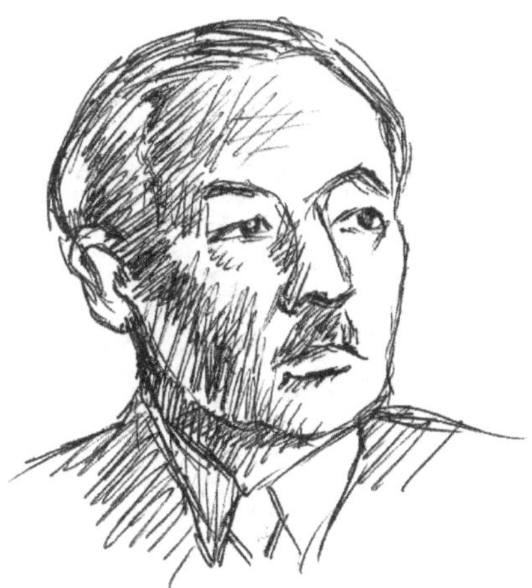

Fig. 2 Han van Meegeren (1889–1947)

The notorious Dutch forger Han van Meegeren (Fig. 2) fooled experts by producing works during World War II that he attributed to Vermeer. These forgeries were cleverly designed to fit the predictions of Bredius, the most acclaimed Dutch art expert of the time. Bredius had said that unknown Vermeers were still to be discovered, that they would have religious connotations and be influenced by Italian art.

It is conceivable that a subconscious desire for his prophecies to come true convinced Bredius that the first Vermeer forgery presented to him *Christ and the Disciples at Emmaus*, which fits all his predictions, was an original masterpiece. Without any close check on provenance, Bredius did not hesitate to declare it authentic.[6] His certification encouraged van Meegeren to forge five more Vermeers using the same approach.

Forged old museum certificates affixed at the back of the painting may also be a way of convincing the curator or gallery owner of the authenticity of the presented art work. The forger might support his claim by a credible story involving family relations, historical events and even forged photographs.

It was however John Drewe (Fig. 3) who exhibited the greatest ingenuity and inventiveness in faking provenance resulting in more than 200 forged works flooding the market between 1986 and 1995.

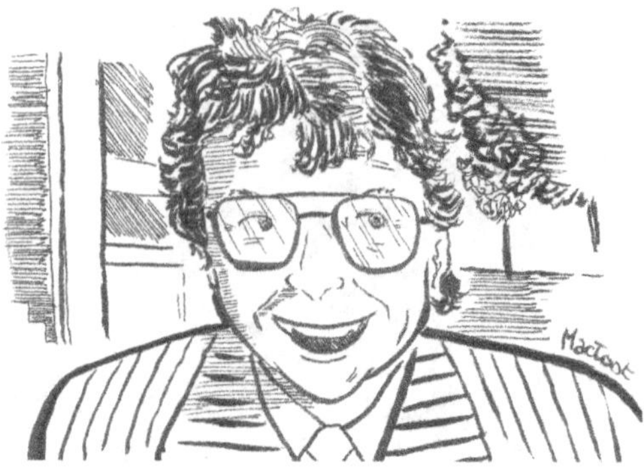

Fig. 3 John Drewe (b. 1948–)

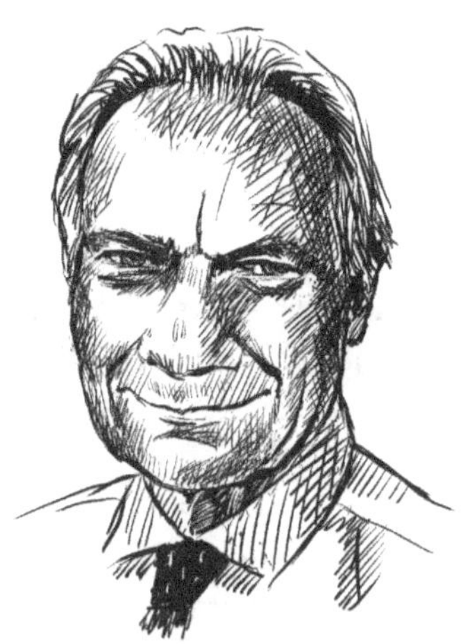

Fig. 4 John Myatt (b. 1945–)

He had identified John Myatt (Fig. 4), a skilled painter who was desperate for money, and convinced him to sell his copies of masters (Matisse, Braque, Giacometti, etc.) as genuine paintings through Christie's and Sotheby's. Drewe fabricated receipts of old purchases, infiltrated the art archives of the Tate Gallery and changed the provenance of authentic paintings to establish a convincing origin to Myatt's forgeries. In an unprecedented scale of corruption, he created catalogues of invented exhibitions and ginned up records of restoration for some of the copies.

Myatt and Drewe were finally brought to trial in September of 1998 and, six months later, convicted to one and six years of prison, respectively, for conspiracy to defraud. Today, Myatt still paints, but sells legitimate copies to fans around the world.[7,8]

Journalist Peter Landesman captures the special irony:

> *Alan Bowness, former head of the Tate and son-in-law of Ben Nicholson, was fooled into authenticating two of Myatt's fake Nicholsons, not because the pictures were good — in fact, the general consensus was that they were unimpressive at best — but because the provenance was flawless.*[9]

Cleverly concocted schemes convinced museums, auction houses and galleries of the provenance of forgeries by the notorious British forger Shaun Greenhalgh (Fig. 5) who produced a wide spectrum of forgeries ranging from Ancient Egyptian sculptures to Assyrian stone reliefs and 19th century watercolours. In a scam that covered a 17-year period, his octogenarian wheelchair-bound father George, acting as an

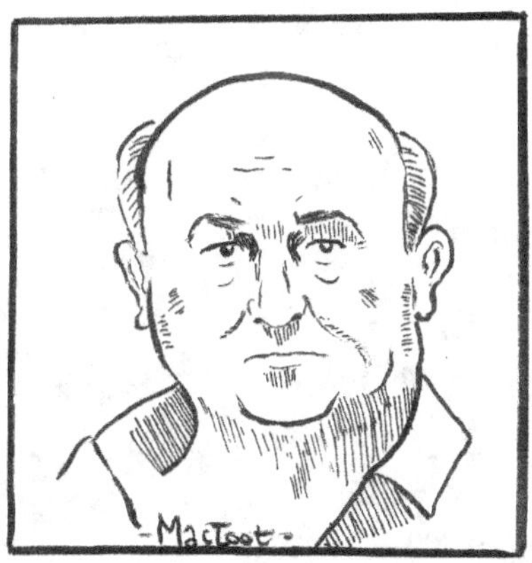

Fig. 5 Shaun Greenhalgh (b. 1960–)

angelic invalid, convincingly carried out most of the transactions together with his wife Olive.

The family would inspect art catalogues and identify missing objects for Greenhalgh to imitate. They would then present the forged objects to the experts as family heirlooms that "might be worth a bob or two".[10] A hint would cleverly be dropped as to where the object came from, followed by a subtle clue that would direct the experts to the provenance of what would have been the authentic art work. It would take no time for the experts to identify the forged object as that alluded to in the provenance, leading to its subsequent authentication.[11]

The most unusual plan to faking provenance was that encountered with New York art dealer Ely Sakhai, who bought a number of authentic paintings by art masters like Gauguin (Fig. 6), Chagall, and Renoir. As related by Charney,[11] Sakhai ensured that these "were not renowned works, and were purchased for prices considered on the low end for fine art" thus avoiding very close scrutiny. He then hired artists to make forged copies of the artworks, and to render these more convincing asked that the markings on the back of

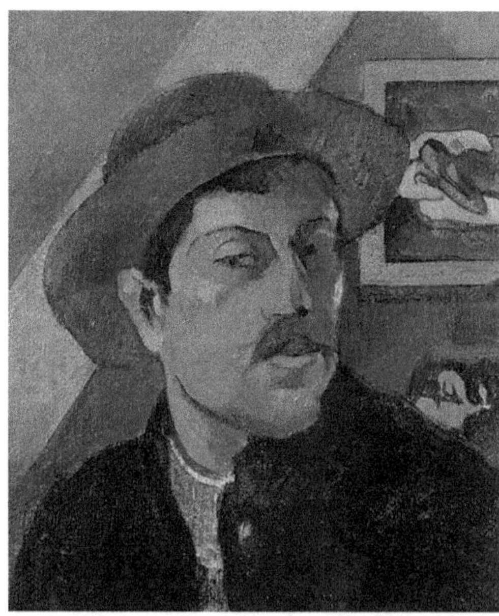

Fig. 6 Paul Gauguin (1848–1903)

the original canvases and frames be reproduced. The legitimate provenance and letters of authenticity that he had obtained for the genuine artworks, would now be attached to the copies[11] and Sakhai would then cleverly seek new letters of authenticity after resubmitting the originals for expert analysis. The copies were sold in Asia, whereas the originals were put forward by Sakhai for sale at auction houses of good repute in Europe and the US.

The whole scam, covering a 15-year period, was finally uncovered when in May 2000 Christie's and Sotheby's released their Spring catalogues and realised, to their great surprise, that they were offering the same Paul Gauguin painting *Vase de Fleurs (Lilas)*. Christie's had received their painting from a gallery in Tokyo, while Sotheby's had received theirs from Sakhai. Sylvie Crussard, a Gauguin connoisseur, later identified the Christie's painting as the forgery.

Forgers may add a signature to give more credibility to the artwork. Curators generally tend to be convinced when presented with a signed painting; however, a cautious curator might have the signature checked by relevant experts.

Forged signatures on alleged Rembrandts were found to abound for the simple reason that greedy forgers signed any painting even vaguely resembling a Rembrandt they could get their hands on. This was witnessed in the Ivan Shishkin affair when in 2004 *Landscape with Brook*, an allegedly important work by this 19th century Russian realist, was withdrawn at the last minute from sale by Sotheby's upon the discovery of a forged signature.[12]

Determining provenance can be a daunting task. As pointed out by the Princeton University Art museum:

> *It is difficult to determine the complete provenance of any work of art. Many objects are bought and sold anonymously; past owners die without disclosing where they obtained the works in their collections; dealers do not always make known the sources of their holdings; and the records of dealers and auction houses are frequently lost or destroyed. For all these reasons, it is rare to have a work of art without some kind of gap in its provenance.*[13]

Curators and gallery owners will therefore seek the help of the expert connoisseur and of the art historian. In recent years, they have also frequently resorted to the scientist.

The Connoisseur and the Art Historian

When the late German art historian Erwin Panofsky compared both the art historian and the connoisseur in art, he expressed it as follows: "The connoisseur might be defined as a laconic art historian, and the art historian as a loquacious connoisseur."[14]

To the acclaimed late art historian Max Friedländer, *connoisseurship* was more of an instinctive exercise, and identifying himself as a connoisseur he wrote:

> *The way in which an intuitive verdict is reached can, from the nature of things, only be described inadequately. A picture is shown to me. I glance at it, and declare it to be a work by Memling, without having proceeded to an examination of its full complexity of artistic form.*[15]

While the connoisseur will judge the style and technique of a particular artwork mostly on the basis of experience, the art historian will additionally bring to bear circumstantial and historical evidence surrounding the artwork. In the present text, both terms connoisseur and art historian will be used interchangeably.

Stylistic and Morellian Analysis

The art historian will make a stylistic analysis to determine if the style and brushwork match those of the artist to which the artwork has been attributed. This will entail comparing the style of the presented artwork with the artist's known body of work as well as with the period style. Clues might be provided by the brushwork, quality of line and colour, composition of the painting and by the historical and relative chronology of the artists' own development. A clever forger will adequately grasp the artist's style and know how to properly emulate the painter's or the period's brush technique. In the case of signed paintings, art historians may also ascertain whether or not the signature is harmonious with the composition of the artwork(s).

A good connoisseur will frequently recognise a forgery even before it is subjected to scientific analysis by identifying the forger's subconscious introduction of a detail that reflects personal style or of an element that is anachronistic with the artist's period. Such was the case in a painting at the National Gallery London depicting a group portrait with the very successful condottieri of the Italian Renaissance, Federico da Montefeltro. Originally attributed to a 15th century master, it contained depictions of anachronistic early 20th century garments.[16]

Morellian analysis, a technique introduced by the physician and art collector Giovanni Morelli (Fig. 7)[17] became popular in the mid-20th century amongst various groups of art historians. It is based on the creation and mapping of formulae describing repeated stylistic details in the artwork and reflecting the particular approach of the artist in creating small features such as ears, eyes, collars, etc. Morelli's study of anatomy might have contributed to his focusing on such details of the human body.

This method can be compared to handwriting analysis and can be understood through Morelli's own words on Botticelli:

> *Among Sandro Botticelli's characteristic forms I will mention the hand, with bony fingers — not beautiful, but always full of life; the nails which as you*

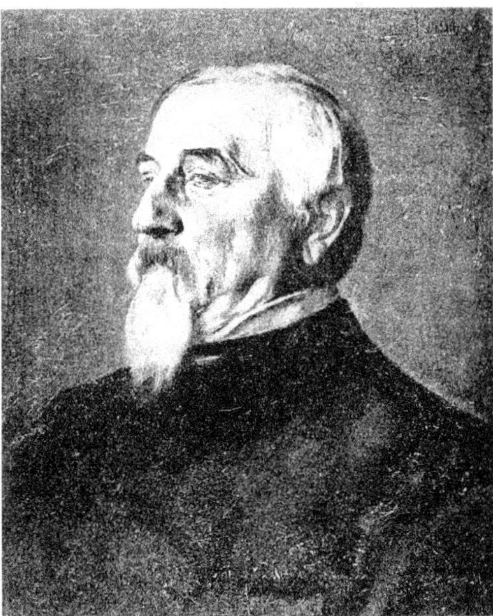

Fig. 7 Self-engraving of Giovanni Morelli (1816–1891)

perceive in the thumb here, are square with black outlines, and the short nose, with dilated nostrils which you can see in Botticelli's celebrated and undisputed work — The Calumny of Apelles.[18]

He also put his case clearly as follows:

As most men who speak or write have verbal habits and use their favorite words or phrases unvoluntarily and sometimes most inappropriately, so almost every painter has his own peculiarities which escape from him without being aware of them.[18]

In recent years, however, reservations have been made on Morellian analysis; some art historians, while recognising that the strength of Morelli's method is its *specificity*, criticise it as being narrowly focused, primarily concentrating on the works while dismissing scientific investigation and other aspects of art historical analysis.[19,20]

Others note that "The Morellian method [...] was swept away by the realization that old masters had workshops or schools who tried to imitate their 'signatures'"[21] i.e. their characteristic approaches in the depiction of certain details.

Some connoisseurs nevertheless still opt to use Morellian analysis, albeit in conjunction with other complementary techniques. Indeed, it might be argued that if today the identification of signature elements is supported by scientific analysis and by additional circumstantial evidence precluding imitation, then these elements are consequently identified as autograph.

The Forger

Sandro Botticelli (1445–1510)

An Irresistible Temptation

Sandro Botticelli (Fig. 8) experienced great artistic acclaim during his early life and middle years. Enthusiastically praised by the renowned mathematician Luca Pacioli, by the famous poet Ugolino Verino[1] and by many contemporaries, his work attracted the attention of the Medici family and earned him their patronage as well as the ultimate honour of being charged to take part in the decoration of the Sistine Chapel. During these years, Botticelli had made himself a name as one of the best painters of his time.[1]

His good fortune would not last forever; with Michelangelo and Leonardo appearing on the artistic scene and with the establishment of the High Renaissance style,[2] his work was slowly pushed away from public attention. In addition, his embracing of the ideas of Girolamo Savonarola, the puritan and fanatic monk who later became the moral dictator of the city of Florence, did not play in Botticelli's favour.

Under the spell of the latter's impassionate ranting, Botticelli, in an extreme act of loyalty to the Dominican Friar, destroyed some of his own work, which he came to perceive as profane and sacrilegious.[3]

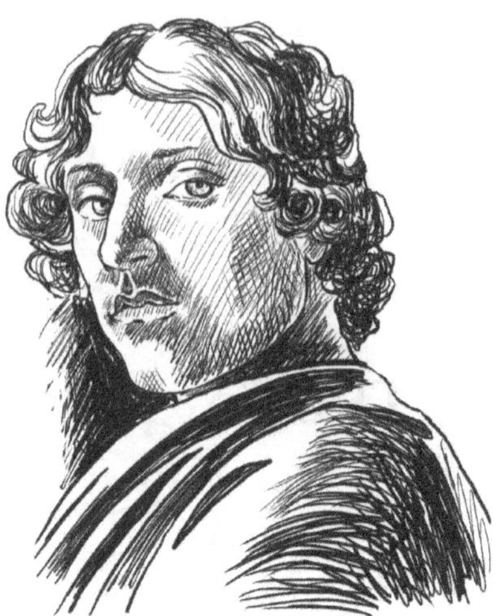

Fig. 8 Sandro Botticelli (1445–1510)

Savoranola had commanded that:

> *Immodest figures should not be painted, lest children be corrupted by the sight. What shall I say to you, ye Christian painters, who expose half nude figures to the eye? But ye who possess such paintings, destroy them or paint them over and ye will then do work pleasing to God and the Blessed Virgin.*[3]

A great decline in Botticelli's artistic career would ensue, he died a crippled man with a minimum number of artistic commissions and was subsequently condemned to oblivion.

His work would only re-enter public consciousness in the late 19th century with full recognition of his prominent role in the Italian Renaissance. It is against the backdrop of such an artistic rebirth that the attention of the notorious *fresco forger*,[4] Umberto Giunti (1886–1970) (Fig. 9) must have been drawn to Botticelli. A teacher at the Institute of Fine Arts in Siena and a picture restorer with the reputation of being a skilful copyist of works by master artists, it is no wonder that copying or creating a work in the style of the newly recognized and famous Botticelli may have appealed to Giunti.

As a skilled craftsman Giunti, along with his master Icilio Frederico Joni, was already engaged in producing convincing reproductions of

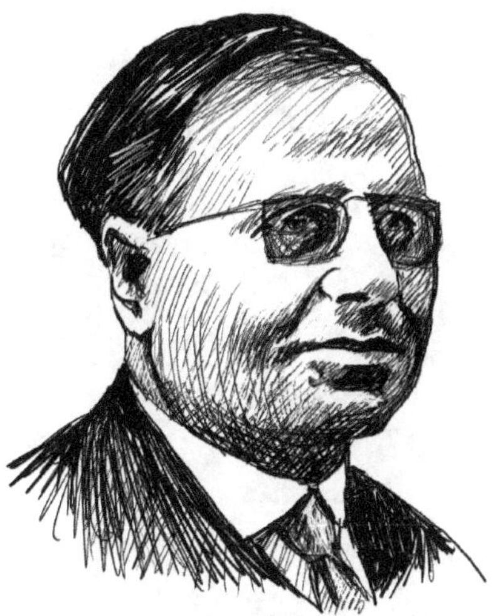

Fig. 9 Umberto Giunti (1886–1970)

works dating from the Middle ages and the Renaissance.[5] As indicated by Joni's memoirs, they operated in such an open manner that it would be difficult to regard such activities as secret.[6] According to the renowned Siennese art historian Gianni Mazzoni:

> *At one stage an entire ring of top forgers was turning out industrial numbers of high quality Old Master paintings in Siena, with the complicity of art experts and dealers.*[6]

It is unclear whether Giunti's copies were initially sold as legitimate copies or presented as genuine artifacts.[5] What seems to be beyond doubt, as revealed by close inspection and scientific analysis of the *Madonna of the Veil*, is that Giunti created this painting, in the style of Botticelli, with a definite intent to deceive (Fig. 10).

Clementina Massimo Giovanelli would later confide to Richard Owen[6] that her uncle Giunti had chosen her as a model for this forgery:

> *I didn't even realize he was using me as a model for the Madonna. I just know he painted beautiful things [...] I had blue eyes and blonde hair, the Botticelli ideal [...] I had no idea my uncle was forging Renaissance paintings to sell on the art market.*[6]

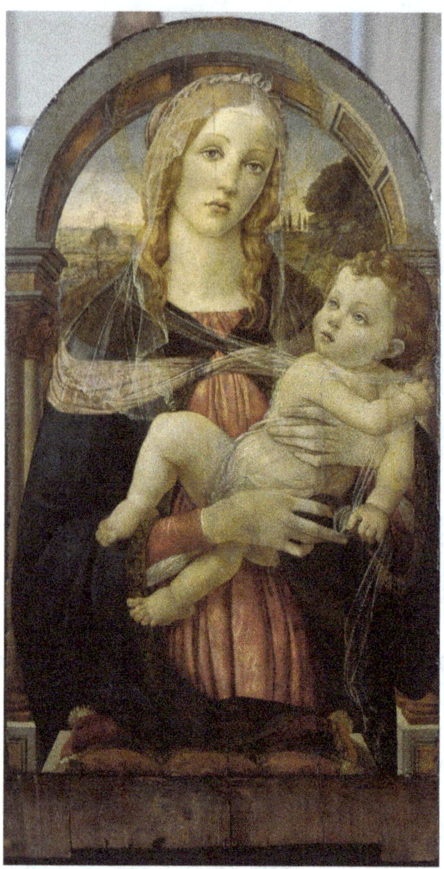

Fig. 10 *Madonna of the Veil*

Scholars believe that Giunti created the *Madonna of the Veil* between 1920 and 1929. In 1930, it was acclaimed by Robert Fry, the English painter and critic, as a beautiful rendition from the hands of the master[7] and attracted the attention of art collector Viscount Lee of Fareham who bought it for $25,000 from a Milanese Lawyer and art dealer.[6]

At a time when Italian primitives had become so very popular in Italy, with demand almost exceeding supply,[5] it is not surprising that Lord Lee would regard this work as one of his prized acquisitions.

In 1947, after his death, the painting went to The Courtauld Gallery in London, with the Medici Society referring to it as "a superb composition of the greatest of all the Florentine painters".[7]

The work had been cleverly conceived by Giunti who masterfully combined elements from several Botticelli paintings into a new and convincing composition. Choosing a theme typical of the master, that of a melancholic Madonna holding the infant Christ, Giunti ensured that the work would preserve the gentleness and peaceful composure of early Botticellis.[8] Today, Madonna's face covered by a subtle translucent veil might appear to the discerning eye as overly tender and romantic when compared with a genuine Botticelli.

Kenneth Clark, the director of the National Gallery from 1934, noted that the Virgin had a great resemblance to Jean Harlow, the attractive American film star of the 1930s. Could Giunti have been inspired by some of the characteristics of this silent cinema star[9] while using his niece as a basis for his production?

Clark's questioning of the attribution to Botticelli led to a closer inspection and investigation of the painting; these would later confirm the legitimacy of his doubts.

It emerged that the Madonna's robe was painted using a Prussian blue pigment, which was only in use in the early 18th century. Analysis by EDXRF (page 151), carried out in 1994, indicated the presence of anachronistic 19th century pigments including cobalt blue, zinc chromate and opaque green chromium oxide,[10] the latter was only available in 1862. Furthermore using an umber pigment, Giunti cleverly depicted the foliage in a brown colour as opposed to green to suggest discoloration due to the passage of time.[7]

Visual examination under a hand-held lens[7] also revealed that the pigments were fine-grained, and not of a coarser kind, as would be expected in the case of 15th century hand-ground pigments. The wooden panel used as backing for the painting was shown by X-ray radiography (page 151)[7] to contain artificial wood worm holes, drilled into the surface, to simulate ageing.[6,8] Finally, examination by optical microscopy (page 135) showed a black outline to the Madonna's lips, in contrast to Botticelli's custom of using a madder lake pigment as a contour for the lips.[7] Here, one can surmise that Giunti may have been influenced by the dark lines around Harlow's lips featured on posters of the actress and subconsciously introduced this atypical element into his creation. As expressed by Wullshlager: "Giunti was highly accomplished but he could not evade the unconscious imprint of his own times."[8]

Today the Madonna of the Veil is a recognized Botticelli forgery at The Courtauld Gallery in London, where it is correctly displayed as a work attributed to Umberto Giunti (1886–1970) and created between 1920 and 1929.

Heinrich Campendonk (1889–1957)

Beltracchi, Germany's Master Forger

Wolfgang Beltracchi (born Wolfgang Fischer in 1951), considered by some as the Robin Hood of Art[1] and by others as Germany's Master Forger,[2] would rattle the art world (Fig. 13).

This notorious self-described hippy[3] had defrauded scores of collectors in one of the greatest art scandals in post-war Germany. The actor Steve Martin paid $860,000 for a forged Campendonk, whereas the Belgian financier and hedge fund manager Pierre Lagrange filed a suit against the Knoedler Gallery after his purchase of a forged Pollock for $17 million. The renowned art dealer Richard Feigen was caught unaware when acting as an agent in the sale of a forged Max Ernst, and the wealthy Dutch collector William Cordia bought a forged painting alleged to be the creation of the Dutch fauvist artist Kees van Dongen for $3.8 million. These are but a few of the numerous forgeries

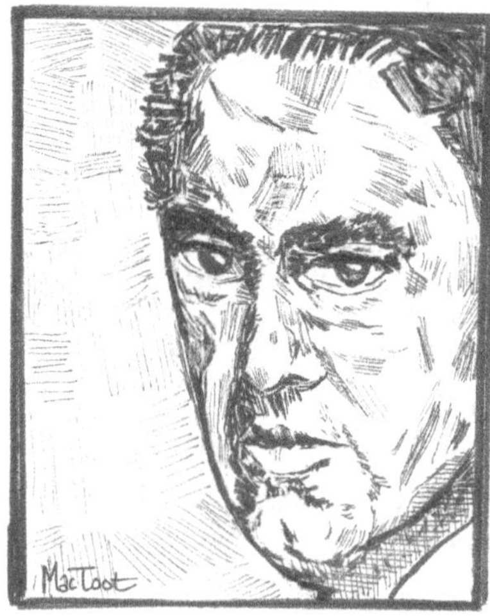

Fig. 11 Heinrich Campendonk (1889–1957)

which were traced back to Beltracchi,[3–5] one of the sharpest, most gifted art forgers of modern times.

Beltracchi sold the forgeries as original works, having masterfully created a convincing account on provenance. He claimed that the works could be traced back to the famous Cologne businessman Werner Jaeger who died in 1992 and even invented the existence of a *Jaeger Collection* allegedly purchased at bargain price from the famous Jewish art dealer Alfred Flechtheim who had fled the Nazis in 1933 and lived in exile in Paris.

The credibility of his story was boosted by the fact that Werner Jaeger also happened to be the grandfather of two of his accomplices, his wife and her sister, who both claimed they had inherited the collection from their grandfather (Fig. 12). Little room was left for doubt when the famous art expert and one of the leading authorities on Max Ernst, Werner Spies, issued certificates of authenticity for a large number of the forgeries. Spies had also been fooled by the story of the two sisters and by the fraudulent Flechtheim labels on the back of the paintings. Thanks to all of this, Beltracchi could continue, unhindered, with the production of his highly deceptive forgeries.

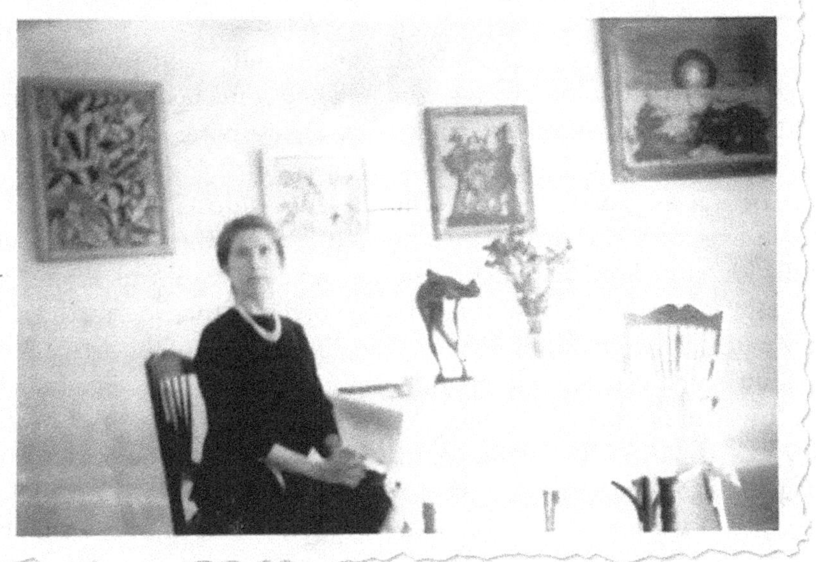

Fig. 12 Doctored picture of Helene Beltracchi with some of the forgeries

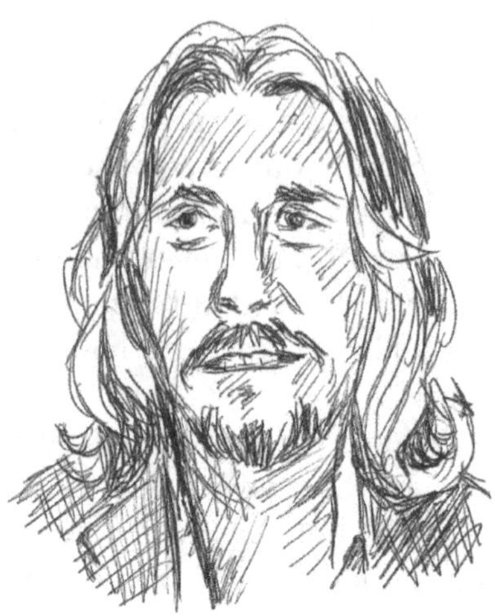

Fig. 13 Wolfgang Beltracchi (b. 1951–)

He meticulously chose materials and pigments in his forged artworks, carefully bought old frames and canvases at flea markets[6] and executed his whole scam with "Mozartian perfection".[7] When exposed, he confessed to 14 counterfeits but seemingly forged a much greater number.

No matter how painstaking the forger's efforts at producing the *perfect forgery*, the time always comes when a mistake is unwittingly made. In forging a Campendonk (Fig. 11) painting *Red Picture with Horses* (Fig. 14), which was sold by Lemperz in Cologne for $3.96 million, Beltracchi had confidently chosen zinc oxide as a white pigment which was part of Campendonk's palette. When the buyer requested authentication from the auction house, no certificate was issued and the painting was then put forward for analysis to the Doerner institute in Munich. Though doubtful about the authenticity of the artwork, the institute's analysts were noncommittal.

It was only through Eastaugh's research carried out at *Art Analysis & Research* (AA&R), using scanning electron microscopy (SEM) (page 138) and energy dispersive X-ray fluorescence (EDXRF) (page 151), that a final verdict was announced — forgery![8] The anachronistic pigment titanium dioxide was identified as an impurity by Eastaugh and could not be attributed to a later restoration as it was found in an underlayer. This initial analysis unveiled the whole scam in the follow-up to which Eastaugh identified several more of Beltracchi's forgeries.

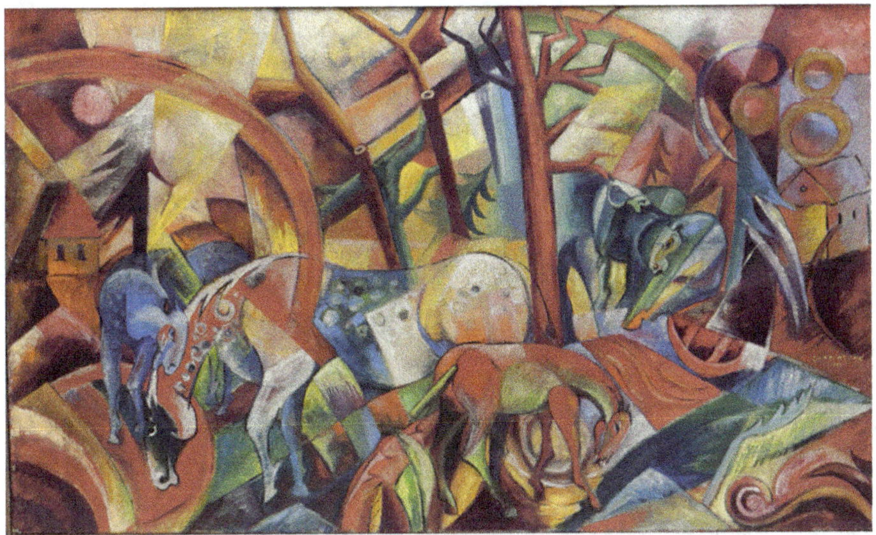

Fig. 14 *Red Picture with Horses* forgery attributed to Campendonk

In spite of the large number of forgeries carried out, on the 27th of October 2011 Beltracchi and his wife were, respectively, sentenced to only six and four years in a German open prison. Those close to the case appeared to be astonished by the lightness of the sentence, including Beltracchi himself who, upon hearing the ruling, joyfully waltzed to jail.

He would later explain that his chief motivation for forgery was to punish what he perceived as the greed and vanity of the art market. He claimed that he wanted to expose the arrogance of the art world, which, according to him, did not really understand the meaning of true art and made arbitrary decisions as to which painting was worth millions and which one was worthless.[9]

Helene Beltracchi was set free in February 2013, less than half-way through her time. Beltracchi's sentence was also reduced and he was released from prison in January 2015. In the meantime, in 2014, striving for further recognition, the couple published their autobiography as well as a compilation of the letters they exchanged while in prison. A film on Beltracchi was also released that same year.

Today, Beltracchi sells and exhibits his own work, and now his paintings all clearly bear his own signature. They are a blend of different styles borrowed from the many modernist painters whose work he had once creatively produced variants of. Ironically, they are in very high demand by an insatiable art market, hungry for works that bear the signature of the famous *Wolfgang Beltracchi.*

The high appeal of his paintings together with the light sentence that he received will undoubtedly encourage other forgers and sadly remind us of the collusion of the art market in the face of unbeatable self-interest.

Entirely Devoid of Creativity?

Beltracchi, explained that he studied very closely the artist he intended to forge, went to museums where the artwork was displayed, visited the place where the artist lived, and tried to observe with the painter's eyes and to penetrate his creative soul.[9] He even claimed that he did not only copy or emulate the style of an artist but wanted to fill the gap in his oeuvre.

It was to his great delight and satisfaction that Dorothea Tanning, Max Ernst's widow and an accomplished artist herself, exclaimed how upon viewing one of his forgeries she felt she was in front of one Ernst's most beautiful paintings.[10]

Beltracchi would boastfully admit, "The trick is to paint a picture that doesn't exist, and yet that fits perfectly into an artist's body of work."[10]

It is in this connection that Meyer[11] and Switzer[12] give us food for thought.

> *The great forgers have not been mere copyists. They have tried — and some have probably succeeded, for we know only their failures — to become so familiar with the style of the master, his way of thinking, that they can actually paint or compose as he did. Thus though their vision is not original, their works of art are, in a sense creations.*

Switzer asks whether a distinction can be made between a copy and a forgery:

> *Or since both are forgeries, between a 'mere copy' of an existing work and an 'original forgery' in which a forger-artist produces a hitherto-unknown work in the style of the great artist — so persuasively that even experts are fooled. The*

first-copying — would strike most of us as merely a certain kind of skill, while the second — actually inhabiting the genius of another and extending it in a novel art-work — is more complicated and more interesting. Here one sees a kind of original artistic vision at work — isn't an artistic vision 'original' to the painter! How can a work be both original and unoriginal at the same time? Clearly it can.

The fine demarcation line in the forger's psyche between the acts of fraud and creativity identified by both Meyer and Switzer provides a fascinating angle of study. Could Beltracchi's brush have been propelled not only by an intention to deceive and mislead, but also by a deep-seated desire to achieve a moment of creative genius?

Marc Chagall (1887–1985)

A Forgery and a Furnace!

In 1992, Martin Lang, a businessman from Leeds, enamoured with art and particularly with the Russian–French pioneer of modern painting Marc Chagall (Fig. 16), bought what he thought was one of the artist's works, *Nude* (Fig. 15), allegedly painted in Belarus in 1909–1910. He paid US $160,000 and had it authenticated by a Russian expert.

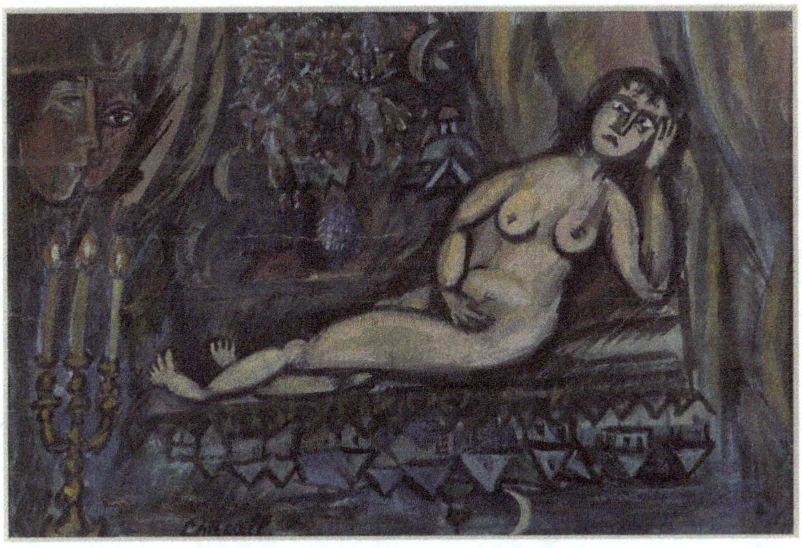

Fig. 15 *Nude* — a Chagall forgery

To confirm its authenticity, Lang lent his painting to the BBC art programme *Fake or Fortune*. The show's presenters Bendor Grosvenor and Fiona Bruce, working with art expert Philip Mould, had it closely analysed and scrutinised.

Professor Robin Clark from UCL was entrusted with carrying out the analyses by Raman microscopy (page 139). Clark identified two key pigments, one blue and one green, which were synthesized at the earliest in 1938 and therefore could not have been available when the artwork was alleged to have been painted.[1]

The painting was later submitted by Martin Lang to the Chagall committee in Paris, formed by a group of experts and headed by Chagall's two granddaughters. The committee also received the results of the analyses (which included the Raman microscopy results). Charged with issuing the final verdict with regard to the authenticity of any Chagall painting which is submitted to it, the committee ruled that it was a forgery.

As pointed out by the head of ArtWatch UK, Michael Daley:

> *Because of the unequivocal nature of those technical findings, Prof. Clark (rightly) observed that the Chagall Committee in Paris, to which the painting was sent, had no option but to confirm the forgery.*[2]

According to French law, discovered fakes are to be destroyed in front of a magistrate. The verdict therefore implied the inevitable destruction of the painting, particularly in light of the previous destruction of two works falsely attributed to Joan Miró in spite of a legal battle with their owners. Expressing disbelief at the verdict, Philip Mould labelled the decision as "barbaric",[3] while Fiona Bruce conveyed both her dismay and concern that no one would ever want to have a painting authenticated by such a committee in the future.[1,3]

The reaction of the makers of *Fake or Fortune* suggests that they were totally oblivious to the dire consequences in the event that the painting was found to be a forgery. Had Lang been aware of such extreme measures, he certainly would have taken into consideration the fact that this particular Chagall had already been exposed as a forgery in Robin Clark's Laboratory at UCL in July 2013.[2]

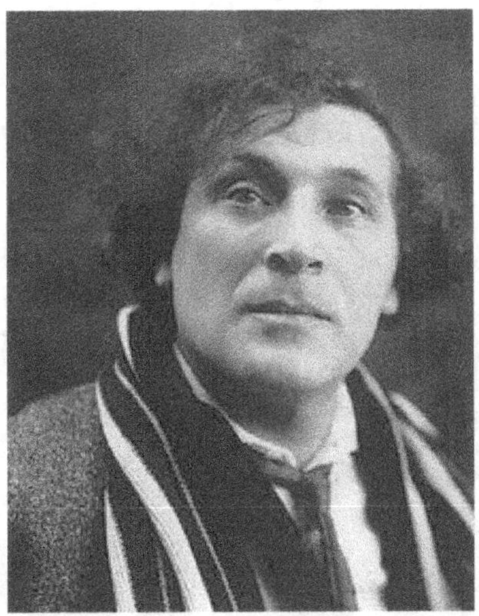

Fig. 16 Marc Chagall (1887–1985)

The Telegraph's chief reporter Robert Mendick wrote that the Chagall affair "marks the culmination of an extraordinary saga, a tale that encompasses the shady underworld of the Russian art market, a journey to Belarus, forensic tests at one of Britain's leading universities and ultimately a French furnace."[4]

Francesco Francia (1450–1517)

The Virgin and Child with an Angel

The German-born industrial chemist Ludwig Mond (1839–1909) had, in his later years, built up an important collection of old-master paintings. At his death, this wealthy collector bequeathed a large proportion of his collection to the National Gallery, London.

Mond had been interested in the Renaissance painter Francesco Francia (1450–1517) who had been apprenticed in his early boyhood to the goldsmith's art[1] but had later tried his hand at painting; this latter endeavour would bring him fame and profit.

One of the paintings left by Mond in 1924 to the National Gallery was *The Virgin and Child with an Angel* (dated 1490) (Fig. 17). He had

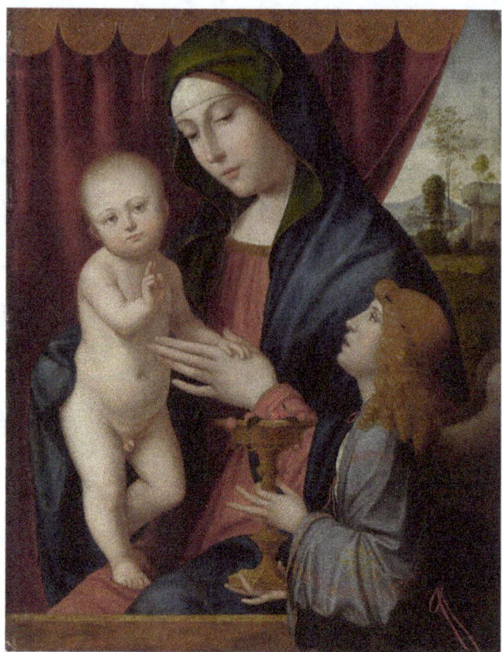

Fig. 17 *The Virgin and Child with an Angel*

allegedly bought this painting attributed to Francia from a Roman dealer in 1893, its earliest provenance being unknown. It was a beautiful artwork and the angel holding an exquisitely crafted golden chalice provided rare testimony to Francia's skill as a goldsmith.

For many years the painting — perceived as one of Francia's earliest works when he had presumably just started his career as a painter — was deemed one of the National Gallery's prized possessions and graced its halls.

Unexpectedly, another identical version belonging to the art dealer Leonard Koetser suddenly surfaced in a 1954 London auction. Quick to take action, the National Gallery scrupulously examined the painting bequeathed by Mond, and in 1955 graciously acknowledged that in great probability theirs was the forged version.

However, it was only in 2009 that the Gallery's painting was confirmed to be a forgery with the other rendition — now in the collection

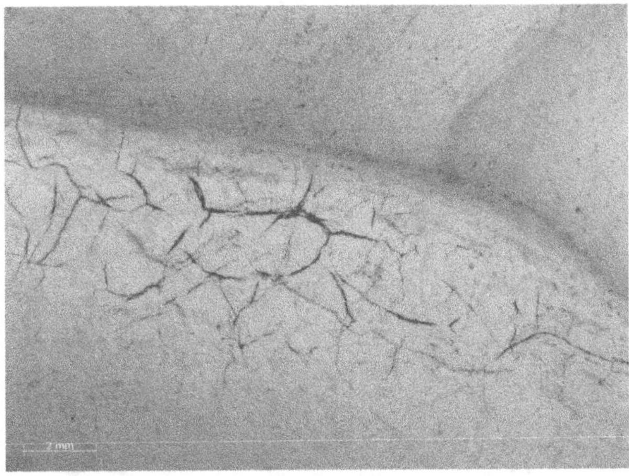

Fig. 18 Faked craquelure observed on *The Virgin and Child with an Angel*

of the Carnegie Museum of Art in Pittsburgh — recognised as the original.[2]

A comparative study carried out on both paintings revealed many inconsistencies in the Gallery version. In the latter case, anachronistic pigments were identified, particularly the red pigment madder revealed by HPLC (page 143), which was only produced in the 19th century. Optical microscopy (page 135) also revealed painted cracks (Fig. 18) on the surface of the 1924 version as well as fine pencil lines in some of the detailed areas, techniques in the latter case totally inconsistent with Renaissance paintings. Finally, infrared reflectography (page 154), indicated that the underdrawing was unusually detailed[2] (Fig. 19) and executed in the uncharacteristic graphite pencil with indications that it had been transferred from a punctured cartoon.

No anachronistic pigments were identified in the 1954 painting, and X-ray images[2] revealed the traditional build-up of the paint layers contrasting with the thin layers observed in the 1924 version. More convincing was the observation of *pentimenti* in the Pittsburgh version confirming the authenticity of the painting.

The very close aesthetic similarity between the two paintings is hard to believe, and yet the Gallery version lost its hold on the viewer.

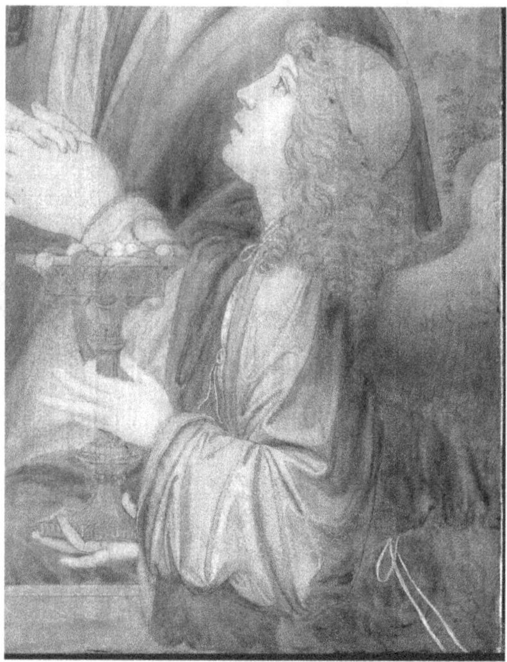

Fig. 19 Reflectogram of the Angel in the National Gallery's painting *The Virgin and Child with an Angel* assumed to have been painted by Francesco Francia (1450–1517)[1]

As eloquently expressed by Jackie Wullschlager:

> *For there is a reason why a downgraded painting not only ceases to command high prices and confer status but also loses its spiritual dimension. An art work's claim to truth lies in fixing a lived moment in time through an artist's mark-making, which is an expression of his hand, eye and mind, and of his era. Denying that truth, fakes loosen our grasp of reality and distort our understanding of the past.*[3]

According to Philosopher Robert Switzer, the arresting quality of an original work comes from a new creative and original execution:

> *In appreciating great art, we want to be a party to a breakthrough of some kind — to a new way of seeing. In these brush-strokes, visible on the canvas, coming from the hand of THIS artist, we feel, we participate in an event of 'originality' that, as artists themselves often attest, seems to transcend the*

individual artist as a person — but it is "through" the artist that something new is coming into the world.[4]

Fernand Léger (1881–1955)

The Bomb Peak Effect Unravels a Forgery

In 1903, Léger applied to the reputable *Ecole des Beaux-Arts* and, sadly, was rejected. He enrolled the same year at the *École nationale supérieure des arts décoratifs* in Paris, and attended the *Ecole des Beaux-Arts* as an auditor for three years. There, he was unofficially mentored by Academicist Jean Léon Gérome, an experience he described as "three empty years".[1,2]

Nothing would impact Léger's work as much as Cézanne's retrospective in 1907 at the *Salon d'Automne in Paris,* after which his art displayed a greater emphasis on the geometry of forms. "Cézanne taught me to love forms and volumes,"[3] he later wrote.

He then adopted a special kind of Cubism where objects were broken down into geometric shapes while maintaining a semblance of three-dimensionality with special focus on cylindrical forms.

By 1913, he created a series of abstract paintings entitled *Contrast of Forms*[4] with the intent of producing a strong effect through the juxtaposition of colours and lines: "I organise the opposition between colors, lines and curves. I set curves against straight lines, patches of colors against plastic forms, pure colors against subtly nuanced shades of gray."[5]

Peggy Guggenheim, the renowned American heiress and former wife of artist Max Ernst, collected a wide spectrum of modern artworks between 1938 and 1946 for her 18th century palace home located in Venice. As of 1951, she would have annual summer displays of her collection which included a painting she believed belonged to Léger's abstract series *Contrast of Forms*.[6,7]

In 1970, she donated her *palazzo* to the Solomon R. Guggenheim Foundation and in 1976 also gifted the foundation with all her works of art. It was then that Douglas Cooper, a Léger scholar, questioned the authenticity of the painting *Contrast of Forms*. The Guggenheim foundation, wary of exhibiting a forgery, decided to hold off on displaying or cataloguing the painting until further investigation.[8] The painting underwent a spectrum of scientific tests with inconclusive results, and the authenticity of the artwork remained a mystery for more than forty years.

In an attempt to solve this enigma, physicists in Florence from the Istituto Nazionale di Fisica Nucleare carried out a crucial and innovative analysis on a minute unpainted edge of the canvas. Using Accelerator Mass Spectrometry (AMS) (page 146), they measured the concentration of radioactive carbon 14 (^{14}C), a technique which would allow them to find out when the cotton used in the canvas had been cut.[8,9]

Such a test is based on the fact that nuclear tests carried out in the 1950s and 1960s increased the level of ^{14}C by almost 100%, resulting in what is referred to as the "bomb peak". The analysis by AMS revealed that the painting was produced in 1959, four years after Fernand Léger's death in 1955,[9] confirming it as a forgery.

Philip Ryland, director of the Peggy Guggenheim Collection made the following statement:

> *After about forty years of doubt surrounding the authenticity of this painting, I am relieved that thanks to the application of innovative scientific techniques, the cloud of uncertainty has at last been lifted and Douglas Cooper's connoisseurship vindicated.*[8]

Old Masters

The Cranach Forgery that Rocked the Art World

> *This is the biggest art scandal in a century. There has been nothing like this since the 'early Vermeer' scandal of the 1940s.*[1]

Venus (Fig. 20) a beautiful rendition of the ancient Roman Goddess of love and beauty, would unleash one the biggest scandals the art world has witnessed to date. The origin of this gripping narrative can be traced back to March 4, 2016, when, in an unexpected move,[2] French authorities seized the *Venus* attributed to German Renaissance Master Lucas Cranach the Elder (1472–1553).

At that time, the artwork was part of an exhibition of the collection of the Prince of Liechtenstein at the Caumont Centre d'Art in Aix-en-Provence. Caught by surprise, the Prince's lawyer expressed his employer's great displeasure by this extreme measure "despite the Prince's long and close collaboration with France's leading cultural institutions, including loans of major works."[2]

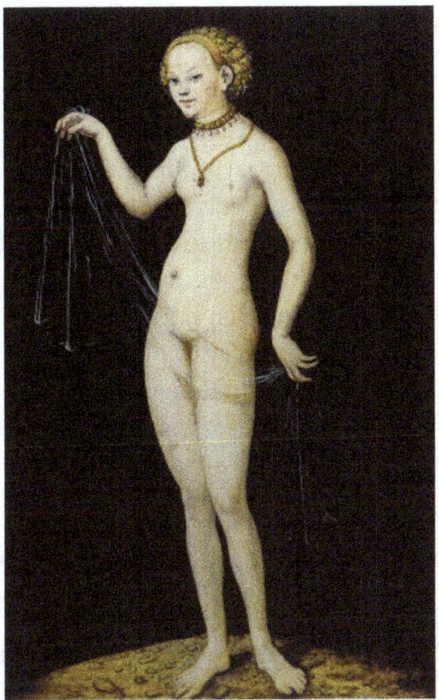

Fig. 20 *Venus* forgery attributed to Lucas Cranach the Elder

French lawyer Corinne Hershkovitch, astonished by such an action, declared:

> *"It is quite unusual for a judge to order the seizure of a work of art, one month and a half after it has been exhibited, and belonging to a European head of State."*[3]

As it emerged, an anonymous letter[4,5] had alerted the French authorities to the matter, led to the confiscation of the artwork and to a criminal enquiry into suspected forgeries created by what appeared to be a highly sophisticated forger.

The whole case would have gone unnoticed, but for the perceptive reporting of Vincent Noce an investigative journalist[2] who immediately broke the news in *The Art Newspaper*. Noce reported that in March 2013, Konrad Bernheimer, the renowned dealer at the Colnaghi Gallery, bought the Cranach for €3.2 million in Brussels from a financier based in Paris. Bernheimer was confident of the authenticity of the work, with three prominent experts confirming it as genuine: Werner Schade, Bodo Brinkmann and Dieter Koepplin. Three months later, the painting was bought by the Prince of Liechtenstein for €7 million to Bernheimer's great satisfaction as only a year earlier it had been put forward for sale and declined by auction houses such as Christie's and Sotheby's.

Analysis of the *Venus* and the detection of artificially-aged paint on a panel created 200 years too late for the German Renaissance painter would later justify the suspicion of French authorities,[6] confirming that is was indeed a forgery.

A closer investigation of the provenance of the Cranach revealed that its origin could be traced back to a collector/dealer by the name of Giuliano Ruffini who lived in Paris.[7]

Portrait of a Man: *Next Link on the Chain of Deception*

A mysterious painting of a man dressed in black, attributed to Frans Hals (1582–1666) was sold by Sotheby's for around $10 million to Richard Hedreen, a Seattle-based collector[7,8] in 2011.

Portrait of a Man had been supplied by Mark Weiss's London gallery and had an interesting past trajectory. It had first appeared on the art market in 2008 when dealer Ruffini presented it to Christie's in Paris, indicating that an art expert had alluded to the fact that it might be by Frans Hals or by one of his followers.[9]

The Louvre's chief curator of Dutch and Flemish paintings, Blaise Ducos, was consulted on the matter and gave his enthusiastic approval of the work. According to him, the elegance and dexterity of execution suggested that it had indeed been created by the hands of the great master. The painting was submitted for further scrutiny to the French Museums Restoration and Research Center where it underwent the conventional examinations of the surface craquelure as well as of the underlayers by X-ray radiography and by infrared reflectography (pages 151, 154). Ultraviolet fluorescence imaging (page 136) complemented the retrieved information. A representative of the centre indicated that as the reassuring results from these tests did not seem to warrant any further scrutiny, pigment analysis was not carried out.[9]

The Louvre was so agog over the serendipitous discovery of what was perceived to be a national treasure that an initial offer of

€5 million was made to Christie's, Mr Ruffini's nominal representative.[2] For some reason the deal did not materialise, possibly due either to a shortage of funding or to the lack of documentation on the painting coupled with the fact that Mr Ruffini refused either to *attribute* or *authenticate* the work.[2]

The purported Hals then attracted the attention of Mark Weiss, the dealer and gallery owner, who bought it and sold it in turn through Sotheby's to Hedreen.

A number of Hals experts continued to praise the newly discovered *Portrait of a Man*, with Quentin Buvelot (Senior Curator at the Mauritshuis in The Hague) and Blaise Ducos referring to it, in a 2014 *The Burlington Magazine* article, as "a very important addition to Hals' oeuvre."[10]

But the news that *Venus* was a forgery and the fact that *Portrait of a Man* could also be traced back to the same original provider Giuliano Ruffini did not escape the vigilant Sotheby's. Quick to take action and very concerned with its own reputation, Sotheby's, with the consent of Weiss, informed the buyer that there might be some issues with the authentication of the Hals.[11]

The next step was to have the painting investigated by the renowned conservator and scientist James Martin at the Massachusetts-based company, Orion Analytical.[12,13] Unfortunately Sotheby's worst fears would be realized, as modern anachronistic materials were detected that could not have been used by the artist. The portrait was declared a forgery and Sotheby's cancelled the sale and reimbursed the buyer.

Weiss, however, refused to pay his share of the sale, casting doubt on the results obtained by Orion and claiming that further analyses were needed. Sotheby's on the other hand claimed that the artwork was in the hands of "experts Mr Weiss had instructed for a four month period and was subject to extensive testing by them. Mr Weiss later suggested that additional tests be conducted by a new group of conservators, but Sotheby's concluded that none of these further tests would change its conclusion."[14]

Sotheby's had no choice but to pursue legal action against Weiss and announced:

> *While we always prefer to settle matters without legal action, the sellers have refused to make good on their contractual obligations and we have been left with no other option than to take appropriate action to enforce our rights.*[15]

When questioned on the whole matter, Ruffini responded that he was merely a collector interested in art, had never claimed the authorship nor the authenticity of unsigned works and that his role was simply to contact the dealers.[16]

More Forgeries from the Same Source: Gentileschi (1593–1653) *and Parmigianino* (1503–1540)

A painting representing Saint Jerome was attributed to the circle of the 16th-century artist Parmigianino.[6] It had been examined by Davide Gasparotto, the expert at The Galleria Nazionale di Parma, Italy, who declared his strong belief that it was by the hand of Parmigianino. It was also displayed at the Kunsthistorisches Museum in Vienna, after careful scrutiny and approval by their own experts.[17]

In January 2012, Saint Jerome was sold by Sotheby's New York for $842,000[6,18] and in 2014, the work was proudly exhibited at the Metropolitan Museum in New York.[6]

As the Old Masters forgery scandal slowly emerged and the origin of Saint Jerome was also traced back to the same source, i.e. Ruffini, Sotheby's once again reacted swiftly and submitted the work to scientific investigation. Using micro-FTIR (page 141), James Martin of Orion Analytical identified the anachronistic pigment phthalocyanine green the choice of which attests to the forger's ingenuity, as phthalocyanine could not have been identified by the conventional technique of X-ray fluorescence (page 149), a method routinely used in museums for pigment identification.

The painting was subsequently declared a forgery and, in a *déjà vu* sequence of events, Sotheby's reimbursed the buyer and sued the vendor Leonel de Saint Donat-Pourrières, who refused to reimburse the proceeds of the sale with the auction house again, declaring: "While we would have preferred to settle this matter out of court [...] we have been left with no other option than to pursue legal action."[18]

Gentileschi

A picture, *David Contemplating the Head of Goliath* (Fig. 21), executed on a slab of the semi-precious lapis lazuli, was attributed to Orazio

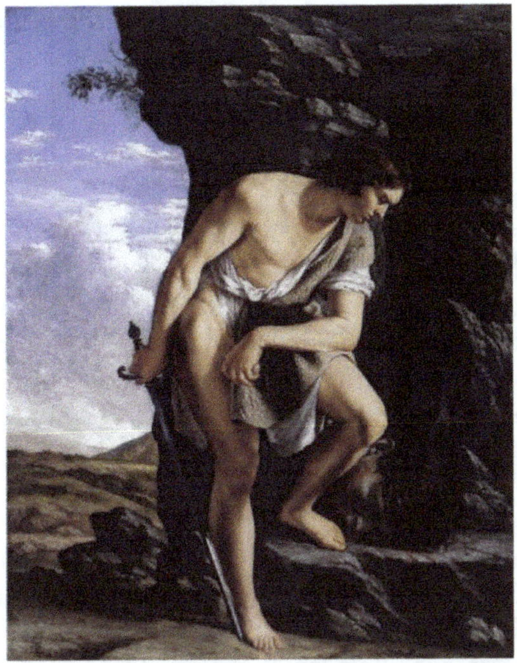

Fig. 21 *David Contemplating the Head of Goliath*

Gentileschi and sold in 2012 by the dealer Mark Weiss. The buyer, a private collector, then loaned it to the National Gallery in London for its 2013 exhibition *Making Colour*.[8]

The National Gallery proudly hailed it as a new discovery,[8] displayed it, and was so convinced of its authenticity that it did not deem the submission of the work to a full scale scientific examination necessary. However, when it soon appeared that the Gentileschi and the Cranach had Guiliano Ruffini as their common point of origin, casting doubt on the authenticity of Gentileshchi's painting, the National Gallery swiftly returned it to its owner. The *David with the Head of Goliath* also appears to be a fake.

The plot thickens

Ruffini claims that a good number of his paintings originate from the collection of André Borie who died in 1971. André Borie, a French civil engineer, worked for the French government on a number of infrastructure projects[13,20] and was a recipient of the Légion d'Honneur. After his death, his daughter Andrée Borie, apparently gave some of her father's works to Ruffini and sold others. What is unusual is that there is no paperwork to ratify *any* of these transactions. Furthermore, there does not seem to be any indication in the historical records that Borie was an art collector.[20]

One cannot help but draw the troubling analogy here with the famous Knoedler and Greenhalgh affairs: just as in the Knoedler affair where the works were purported to have belonged to a certain Mr X, the Ruffini paintings, said to have been part of the André Borie collection, do not have an established provenance. Also, Mr Ruffini's insistence that he never presented a single painting as the real thing[1] but had let the experts decide, reminds us of the *modus operandi* of forger Greenhalgh.

Although Greenhalgh cleverly fabricated a convincing provenance for his forgeries, the works were presented to museums and galleries as family heirlooms, with no claim of attribution or authenticity. Sometimes a small hint would be given at what they might be, but it was left to the experts to decide what they represented.

The difference with Ruffini however, is that when Greenhalgh's house was raided there were, as noted by an astonished Scotland Yard detective: "blocks of stone, a furnace for melting silver on top of the fridge, half-finished and rejected sculptures, a watercolour under the bed."[21]

When Ruffini's estate was raided by French and Italian police, there were no incriminating findings.[9] At present Ruffini seems clear of any charge and the investigation is progressing, with rumours suggesting that as many as 25 Old Masters works[1] may be revealed as forgeries. The plot thickens!

Jackson Pollock (1912–1956)

A Trove of Pollockesque[1] Paintings. Authentic?

Pollock's art was described as "volcanic [...] unpredictable [...] undisciplined",[2] and his seemingly easy technique of pouring or dripping

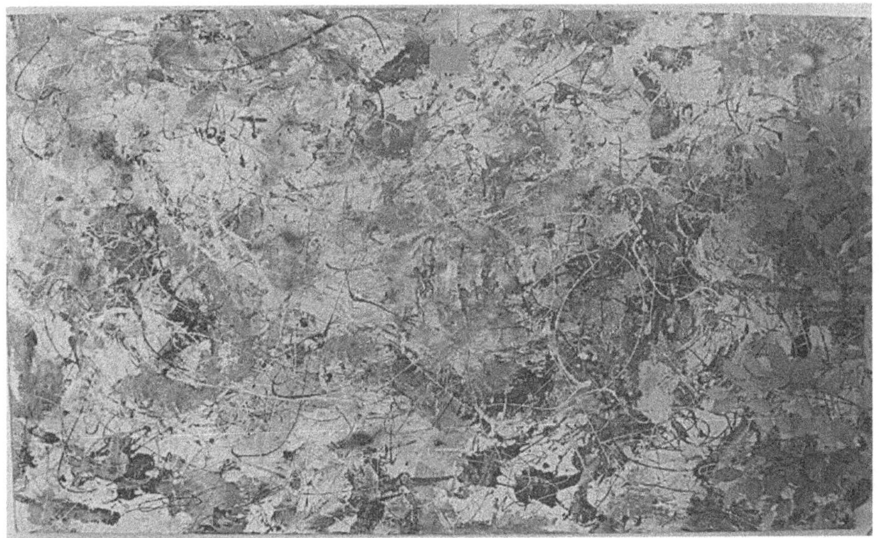

Fig. 22 Pollock style 2009 rendition

paint on to the canvas (drip painting) gave it, in the eyes of forgers, tantalising appeal. As they desperately tried to emulate his abstract expressionist approach, time and again, Jackson Pollock forgeries made headline news.

No news, however, would cause more ripples in the art world than the many brown paper packages that Alex Matter discovered in his parents' attic in 2002. The packages proved to contain 32 *Pollockesque*[1] paintings left by Alex's father photographer Herbert Matter after his death in 1984 and that appeared to closely reflect Pollock's style: "Just to paint … a gesture of liberation from value — political, aesthetic, moral."[3]

Matter called upon the expertise of scientists at the The Harvard University Center for the Technical Study of Modern Art to analyse three of the paintings in 2002.[4]

The research group at Harvard offered their services pro bono, requesting paintings that had not undergone any conservation treatment after their discovery.[4] As their wish could not be fulfilled, the team had to work with three artworks that had already been treated by a conservator.

After concentrating on areas that had definitely not undergone any conservation treatment, the Harvard team used a synergy of scientific techniques to analyse these artworks which included laser ablation inductively coupled plasma mass spectrometry (LA-ICP-MS) (page 144) and pyrolysis gas chromatography mass spectrometry (Py-GC-MS) (page 143) for the analysis of pigments and for the investigation of some of the binding media. All the results were related in a special report.[4]

The team found red paint, which was marketed only a few decades ago, and two pigments — one in an orange paint and another in a brown paint — that only came into the market in 1971 and 1986 respectively. Py-GC-MS also revealed a terpolymer not commercially available before 1970 which was used as a binding medium to produce a silver paint.[4] The presence of all of these pigments discovered after Pollock's death refuted the authenticity of the paintings.

Matter was not inclined to accept the verdict; what could allay his scepticism when Jackson Pollock (Fig. 22) had been such a close family friend and when Pollock expert Ellen Landau had said, "There are too many things about them that are pure Jackson..."?[5] It would prove to be a long and surprisingly controversial affair.

With regard to the anachronistic pigments, Matter argued that Pollock got his supplies from a store in Switzerland run by his father's relatives, the store sold paints only later patented in the United States.

Landau's inclination to view the paintings as authentic (she had also spent many years exploring the relationship between the Pollock and the Matter families) had been challenged by art dealer Eugene Thaw who, along with Francis O'Connor, had drawn up a four-volume listing of Pollock's work (the catalogue raisonné), which was regarded as the most reliable source on authenticity matters.[5]

In 2005, Alex Matter went on to ask expert forensic scientist James Martin to analyse 24 paintings. Martin's laboratory was equipped with state of the art micro-FTIR (page 141), a technique which would play a crucial role in reaching a final verdict on the artworks.[6]

Although Martin's analysis led him to the discovery of pigments that suggested the paintings were inauthentic, he did not initially release any of his findings for fear of libel.[7,8] Later on, when he did announce his results pointing to the identification of anachronistic pigments in the paintings, they were deemed unconvincing. When Alex Matter's lawyer was asked in an interview whether or not his client was prepared to accept the fact that the paintings were inauthentic, he replied, "No. We don't regard Mr Martin's conclusions as reliable."[7] It would gradually however become evident, through a closer look at Martin's results, that his conclusions were unchallengeable and had to be taken very seriously.

Martin described his discovering of the anachronistic Pigment Red 254 in 10 of the paintings as the "'Aha!' moment"[9] and "that it was strong evidence that those pieces were not created by Jackson Pollock"[9] Pigment Red 254, known as 'Ferrari red' because of its being "used on all solid-red Ferrari's from 2000 to 2002",[10] was patented by Ciba-Speciality Chemicals in Basel, Switzerland in 1983, well after Pollock's death. This pigment based on the organic compound called DPP (diketopyrrolopyrrole) was detected by Martin using micro-FTIR when he analysed minute paint chips taken from different layers of the purported Jackson paintings.

The question remains: Who produced these paintings? Could it have been Herbert Matter's wife Mercedes, an artist in her own right? Could it have been her students? Or was it a joint endeavour as an educational and experimental venture into drip painting? No one will ever know!

What is clear however is that in this particular case, the absence of Jackson Pollock's signature renders the classification of the paintings as *inauthentic* more apt than *forgery*.[9]

Authentication

Lucian Freud (1922–2011)

An Artist vs. the Experts: Man in a Black Cravat

I am completely selfish and I only do what I want to do.

Lucian Freud[5]

A cruel lover, a terrible father, a reckless gambler, a control freak and a loner who hated the intrusion of people; Lucian Freud — grandson of famous psycho-analyst Sigmund Freud — was considered to embody all of these traits but also, to be one of the greatest portraitists of the 20th century (Fig. 23).

A self-absorbed man, who referred to Picasso's art as "absolutely poisonous", to German expressionist Max Ernst as a "heavy and stiff" dinner companion,[1] Freud kept relationships in private compartments: to those he liked he was "incredibly sensitive and deeply considerate[2] [...] thoughtful, and devilishly charming",[3] and to those he disliked he could lie and be abusive. It is even said that he sometimes paid criminals to intimidate those who did not meet with his approval.[4]

He submitted his naked subjects stripped bare and unidentifiable by class or social standing to his powerful and ruthless inspection. Giving in to his charisma and magnetism, trapped by his peering and piercing eyes, they were happy to endure the long sittings for him while he redefined figurative art and reshaped the art of portraiture.[6]

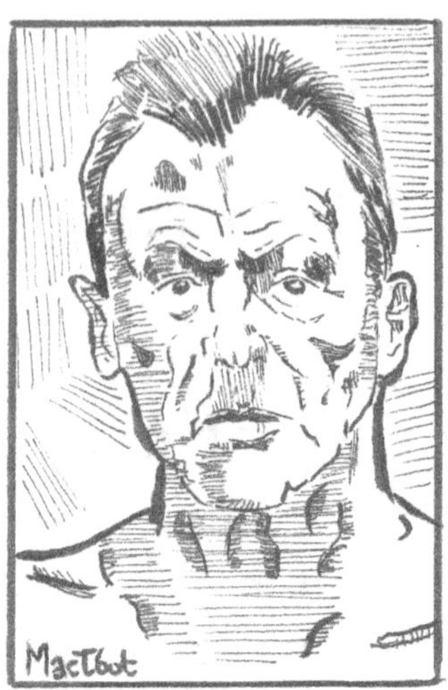

Fig. 23 Lucian Freud (1922–2011)

Freud started painting portraits early on, when still at art school. Later on in his career, these relatively simple executions would evolve into grave and thickly impastoed renderings characterised by deep psychological penetration.

Against the backdrop of such a complex character, it might be easier to understand why Freud adamantly rejected one of the early portraits attributed to him entitled *Man in a Black Cravat*.

The story begins during the Second World War, when artist Wirth-Miller and his boyfriend Richard Chopping unearthed a portrait entitled *Man in a Black Cravat* (Fig. 24) in a barn at the East Anglian School of Painting and Drawing.[7] The couple thought the portrait was an early work by Freud.

A long-running feud between Wirth-Miller and Freud seems to have started whilst training at the East Anglian School where the two artists studied. They reportedly intensely disliked one another with Wirth-Miller drawing lists of "Reasons I hate Lucian Freud" and Freud calling Wirth-Miller: "Worst Miller".[8]

Freud spent years persistently denying that he had painted *Man in a Black Cravat*. In 1985, Christie's put it up for sale as a portrait by Freud (this view being backed by its own experts); the sale was cancelled

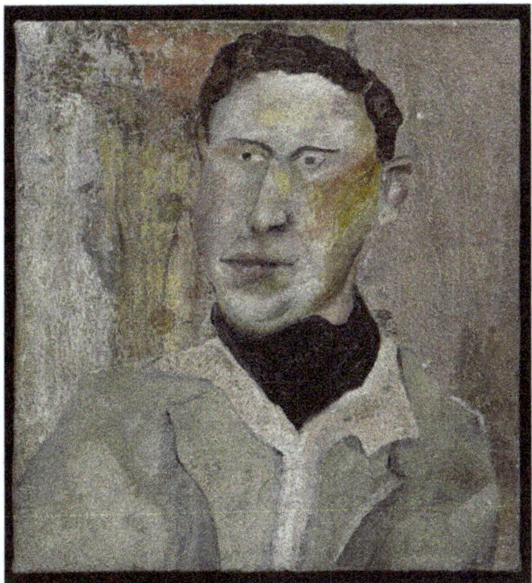

Fig. 24 *Man in a Black Cravat*

however as it was claimed that Freud — still alive at the time — was unable to authenticate it.

The years would pass, and in 1997 Jon Turner, a London-based designer, received the work as a gift from Wirth-Miller and Chopping. He in turn would spend many years trying to authenticate it. He always recalled Wirth-Miller's words who had since died: "I want you to sell this picture as publicly as possible. I want you to humiliate Lucian Freud."[7]

It is then that Turner decided to resort to *Fake or Fortune*, the renowned BBC series examining the provenance and attribution of high-profile artworks. Could the presenters of the programme journalist Fiona Bruce and art historian Philip Mould, together with their specialist art researcher Dr Bendor Grosvenor, address his predicament?

Accepting the challenge, but realising the difficulty of such a daunting task, Fiona exclaimed, "It was a gargantuan task to overturn the reported views of the artist. It was different from anything we'd taken until now."

The first step in the presenters' investigation was to interview novelist Rose Boyt, one of the 14 children Freud acknowledged as his own, who revealed that Jon Turner had approached her in 2006 wondering if she could seek her father's opinion on the portrait. She had declined to do so, claiming to the interviewers that she had then feared he would react violently to the point of putting his fist through it and destroying it.[9] Boyt went on by saying, "He hated the intrusion of people asking: Did you do this or not? I thought if he hadn't identified it in the normal course of things, that meant to because it was stolen, it wasn't by him or he hated it."[9,10]

A breakthrough however was made when Bruce and Mould tracked down the artist's former trusted solicitor Diana Mary Rawstron. The latter found a note of a 2006 telephone conversation with Freud, where he acknowledged that he had started painting the shirt, body, neck and part of the head in the portrait,[10] but that it had been completed by someone else. Rawstron also confided to the presenters that Freud on any query regarding the portrait insisted that her answer be that he was unable to authenticate the painting, without volunteering the added information that he had painted parts of it.[10]

Did Freud act in retaliation to Wirth-Miller because of their lifelong feud? Was it a vengeful act and a means of preventing the sale of the painting? To the BBC presenters, who could only speculate on the answers to such queries, the only way forward was to try to find out if indeed the portrait was the work of a single artist.

In the meantime, further archival investigations carried out by Grosvenor identified a fellow student at the East Anglian School of Painting and Drawing called James Jameson as the subject of the portrait, with 1939 being the most likely date of execution of the work (when Freud was only 17 years old).[10]

Art historian Mould, in a quasi-Morellian approach, noted features in the portrait that were characteristic of Freud: the elongated asymmetric face with its wilful distortion, the off-centre nose, tilted mouth and piercing eyes. These features were compellingly similar to another authenticated early portrait by Freud[10], painted in 1940, which also displayed the same physical distortions and psychological intensity. The brush strokes were comparable as was the choice of colours and the mixing of colours on the faces.

What was now left for Bruce and Mould was to seek the viewpoint of a technical art scientist so as to ascertain once and for all if there had been one executing hand.

Libby Sheldon from University College London was approached for that purpose, and, after carrying out her analyses, she confirmed that the whole portrait seemed to have been done by the same artist. Close examination using polarised light microscopy, EDXRF and cross-section analysis[11] (pages 136, 151) showed that the original wet paint of the shirt overlapped with that of the scarf, the blackish blue pigment of the latter matched the black pigment of the hair and the same nature and range of pigments as well as of their mixtures were applied throughout the portrait and in the same fashion.[10]

The puzzle was almost entirely solved, but for a final ruling on authenticity there needed to be a seal of approval from art experts on Lucian Freud. With great anticipation and some trepidation, a panel of three such experts were consulted: namely William Feaver, James Kirkman and Toby Treves.

To the great delight of Bruce, Mould and Turner, all three experts concluded that Freud had painted the portrait, most probably in 1939. There was however a caveat: one of the members of the panel, Toby Treves, who was in charge of Freud's catalogue raisonné though conceding that the whole figure was painted by a single person expressed strong doubts that the background landscape was also painted by Freud. Accordingly, the painting would not be included in Freud's catalogue raisonné, but only referred to in an appendix of the book.[12]

This long journey into the unknown, with all the "arm wrestling with the words of an artist beyond the grave"[10] that it had entailed, ended up with victory and has finally been settled.

Will Turner fulfill Wirth-Miller's wish and sell the portrait now evaluated at £300,000, as publicly as possible?

Edouard Manet (1832–1883)

Infanta Margarita

Walter McCrone, the leading American microscopist, received a call in 1983 from lawyer and art collector Andrew Brainerd who had purchased an artwork somewhat reminiscent of the *Infanta Margarita* by the great Spanish master Diego Velázquez. The painting, purchased in Amsterdam a few years earlier, depicted a young blond girl, her hair held back with a pink ribbon.[1]

Brainerd subsequently learned that the renowned French realist painter Edouard Manet had been a strong admirer of Velázquez, and had made an oil reproduction of the *Infanta Margarita*. He contacted McCrone, who agreed to examine the painting using polarised light microscopy (page 137).

McCrone's analysis of a dozen differently coloured pigments confirmed that they were all in keeping with a 19th century palette, and that the pigments had *atypical configurations*.[1]

Intrigued, he subsequently managed to obtain microscopic samples from two of Manet's certified paintings — *Spanish Ballet* and *Woman with a Jug* — painted in 1862 and in the period 1859–1862, respectively. His investigation of the paintings by polarised light microscopy revealed that in both paintings, pigments responsible for the white, blue and red colours exhibited the same atypical configurations observed earlier in Brainerd's painting. Furthermore, when the lead-white pigments were analysed, they were found to contain the "same nine trace elements, in practically identical proportions".[1] This could only have been the case if the two paintings had been prepared the same year, from identical production lots as they could not be "more similar if they had been squeezed from the same tube of paint."[2]

These results were the crucial initial steps towards the authentication of Brainerd's painting as Manet's copy after Velázquez's *Infanta Margarita*. Further historical, stylistic and scientific analyses by Boime and Kossolapov[3] in 2003, who used X-ray radiography (page 151), led to the definite attribution of this painting to Manet and the deduction that it had been painted in the early 1860s.

Recognising the critical importance of science for this particular attribution case, Boime and Kossolapov stated:

> *The results of this scrupulous examination ratify what the heart has known all along, and at long last we can state with a comfortable degree of certainty that the outcome of the application of the latest scientific methods to this picture has eradicated whatever reservations specialists may have felt over the years about the two conditions noted above. Although art historians and art experts seem to be nervous about relying too heavily on the application of conservation science for authentication, in a case of this sort, conservation science should be seen as the inevitable and necessary adjunct to sound connoisseurship.*[3]

Mahmoud Said (1897–1964)

Case of Government Alarm

Christie's announcement of a sale to take place on October 31st, 2007, of Mahmoud Said's painting *The Girl with Green Eyes* raised eyebrows and alarm in Egypt's cultural milieu. This artwork was supposed to be hanging in the home of the Egyptian ambassador to the UN in New York. Was the painting on sale a copy of the work or had the authentic version in New York been replaced by a forged copy?

Alerted, the Egyptian government promptly intervened, contacted Interpol and the sale at Christie's was immediately cancelled.[1]

There is no doubt that in the last two decades, a booming market and a surge in demand for Middle Eastern art in general and for modern and contemporary Egyptian art in particular has led to the emergence of a large number of fake artworks. Producing forgeries of Said's paintings, now selling for millions of dollars, would certainly be an attractive prospect for forgers and dishonest art dealers.[2]

Said's art is mesmerising: his brilliant palette brings to life many aspects of Egyptian culture and his compositions are compelling in their juxtaposition of shadow and light. His nudes and female depictions are often somewhat erotic, and his voluptuous depiction of *The Girl with Green Eyes* bears a lifelike three-dimensional quality (Fig. 26).[3]

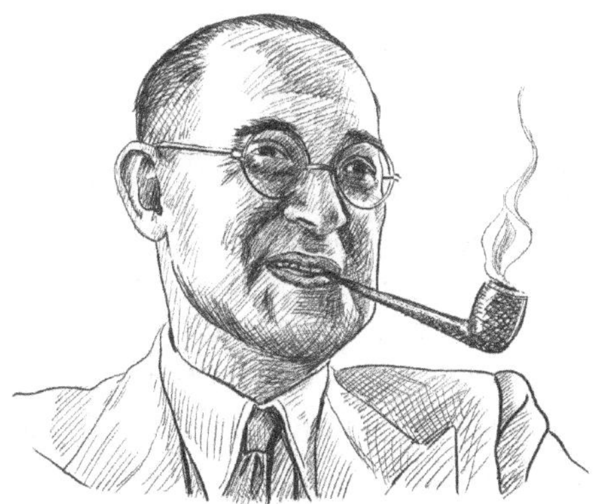

Fig. 25 Mahmoud Said (1897–1964)

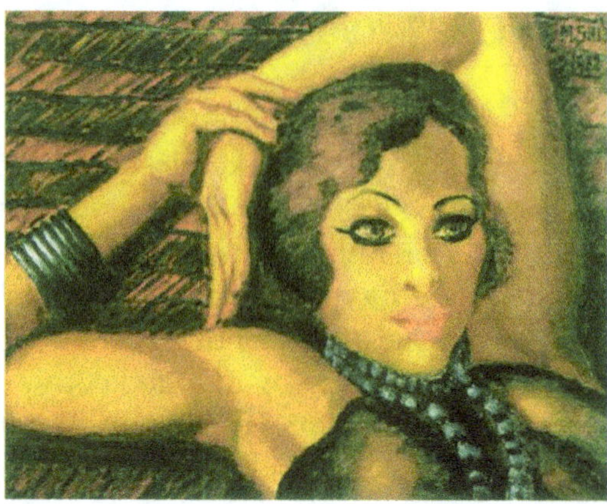

Fig. 26 *The Girl with Green Eyes*, 1931 rendition

Described by his niece, the former Queen Farida of Egypt, as "a quiet, gentle, oppressively timid man,"[4] Mahmoud Said (Fig. 25) was born in Alexandria in 1897 and belonged to a wealthy aristocratic family. His father Mohamed Said Pasha, a former prime minister of Egypt, insisted that his son pursue a career in law. To the whole family, it was unbecoming for a young man with a privileged upbringing as was his to become an artist. He had been educated in some of the best schools in Egypt and was expected to engage in a more serious career.

In a 1983 interview, Queen Farida would describe his family's disapproval of his inclination for art: "They simply could not relate to his art […] The closest they came to showing a sign of approval was when they acknowledged that his work must be good because foreigners seemed to appreciate it."[5]

Fulfilling his father's dream, Said first worked as a lawyer then became a prosecutor and finally a judge in the Cairo, Alexandria and Mansouria governorates. He painted in all of his free time, creating a significant body of work including *The Girl with Green Eyes* (in 1931).

A few years after his father's death, Said set out to fulfil his lifelong dream. He abandoned his law career and, as of 1947, until his death in 1964, became a full-time artist.

The Girl with Green Eyes was acquired by the Egyptian Museum of Modern Art in 1935, and in 1950 it was sent on loan to the Egyptian embassy in Washington DC. Almost a decade later, the painting was transferred to the residence of the permanent representative of Egypt's mission to the UN, where it still hangs to this day.[6]

In this case, to differentiate between the two renditions of *The Girl with Green Eyes*, the most appropriate scientific approach would be to analyse both paintings by Raman microscopy or by FTIR microscopy (pages 139, 141) in the hope of identifying anachronistic elements in either of the two artworks.

Alternatively, it might also be appropriate to look for the presence of the two isotopes (^{137}Cs) and (^{90}Sr) in both paintings. Any artwork, purported to have been created before 1945, as was *The Girl with Green Eyes*, and found to contain these two isotopes would necessarily be forged (page 146).

Fortunately, the committee appointed by the Egyptian government to investigate the affair was in for a pleasant surprise. To everyone's great relief, it emerged that a January 1936 edition of the weekly magazine *La Semaine Egyptienne* indicated the existence of two versions of *The Girl with Green Eyes* referring to one as painting #2 (Collection of the Egyptian Museum of Modern Art in Cairo) and to the other as painting #80 (Charles Terrace Collection). It transpired that, in 1936, Charles Terrace had acquired this second 1932 version, from the artist himself[2] (Fig. 27).

As depicted in Figs. 26 and 27, though very similar, the two creations are not identical.

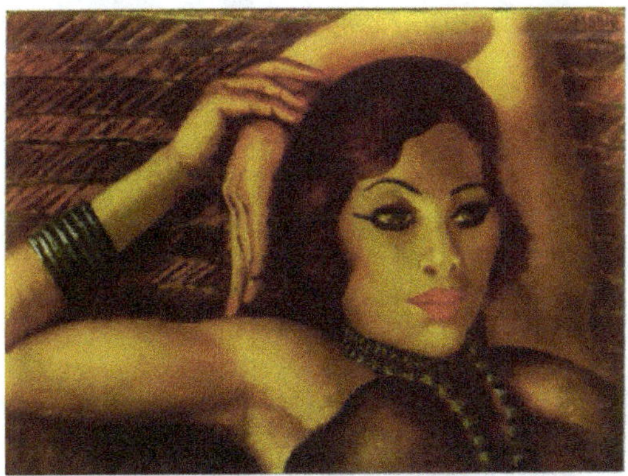

Fig. 27 *The Girl with Green Eyes*, 1932 rendition

Provenance here had provided the key element to *authentication*, and had come to the rescue.

Vincent van Gogh (1853–1890)

Two Paintings and a Set of Letters

Much of what we know today about Vincent van Gogh (Fig. 28) comes from a total of 928 surviving letters, 650 of which were addressed by him to his brother Theo.[1] When Theo died in 1891, his widow Jo van Gogh Bonger compiled the whole set of letters and published them in a first edition in 1917. The letters, originally written in Dutch and French, were later translated into English in 1958 and played a crucial role in the authentication of two of van Gogh's paintings.

Hitherto Unknown Floral Still Life

In one of the letters dated January 22, 1886, the Dutch master wrote to his brother:

> "This week I painted a large thing with two nude torsos — two wrestlers."[1]

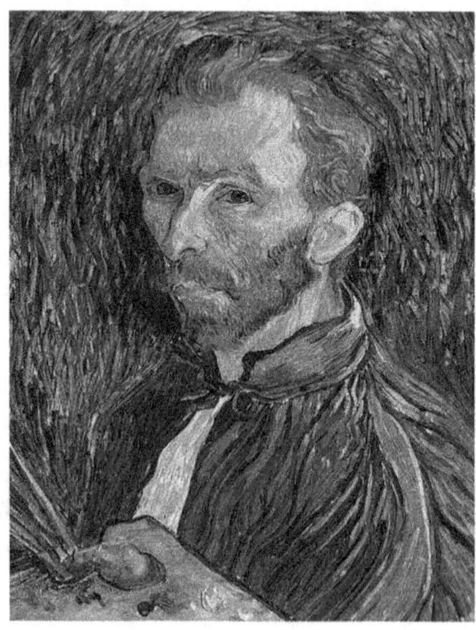

Fig. 28 Vincent van Gogh (1853–1890)

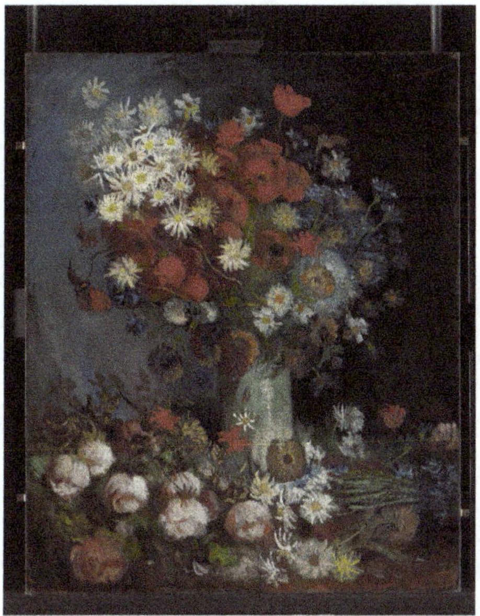

Fig. 29 *Still Life with Meadow Flowers and Roses*

A floral still life entitled *Still Life with Meadow Flowers and Roses* (Fig. 29) acquired in 1974 by the Kröller–Müller Museum in Otterlo near Arnhem was considered uncharacteristic of van Gogh because of the unusual size of its canvas (100 × 80 cm), the position of the signature and the large number of flowers, the latter being an uncommon feature in van Gogh's still lifes.

As van Gogh often reused the canvas of an older painting and created a new artwork on top, in 1998 the above-mentioned still life was analysed by X-ray radiography (page 151). It gave a very unclear underdrawing which was difficult to interpret and in 2003, the museum finally listed the painting as 'Artist: Anonymous' and relegated it to a back room away from public view.

In 2012, in view of the rising recognition of XRF synchrotron imaging (page 152) as an important new tool for the analysis of paintings, the decision was made to revisit the painting and investigate it using the latter technique.

Experiments conducted at the synchrotron radiation facility in Hamburg revealed an extremely clear underdrawing of the two

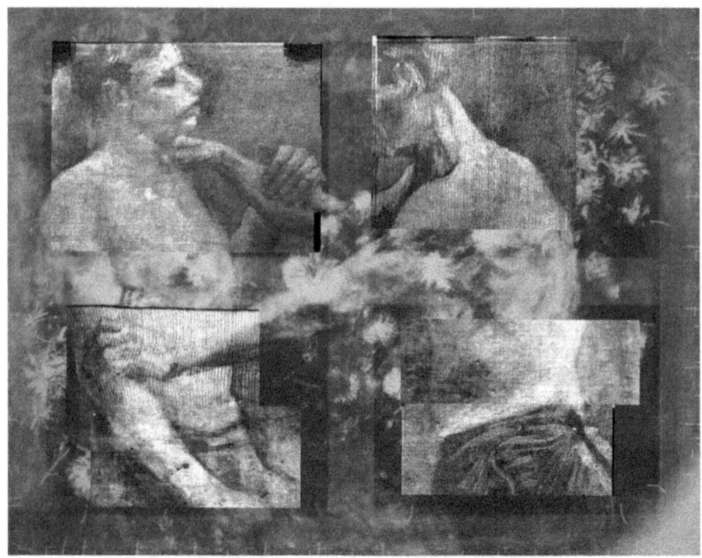

Fig. 30 Underdrawing showing the two wrestlers

wrestlers (Fig. 30) in which van Gogh's characteristic brush strokes could be well recognised. X-ray fluorescence analysis of the floral painting as well as that of the underdrawing indicated pigments in keeping with van Gogh's palette, thus confirming that the painting was indeed the one he had referred to in his letter to his brother Theo.

In March 20, 2012 *Still Life with Meadow Flowers and Roses* was officially and proudly displayed at the Kröller–Müller Museum as an authentic van Gogh.[2]

Sunset at Montmajour: *a Genuine van Gogh*

Yesterday, at sunset, I was on a stony heath where very small, twisted oaks grow, in the background a ruin on the hill, and wheatfields in the valley. It was romantic, it couldn't be more so, à la Monticelli, the sun was pouring its very yellow rays over the bushes and the ground, absolutely a shower of gold. And all the lines were beautiful, the whole scene had a charming nobility. You wouldn't have been at all surprised to see knights and ladies suddenly appear, returning from hunting with hawks, or to hear the voice of an old Provençal troubadour.[3]

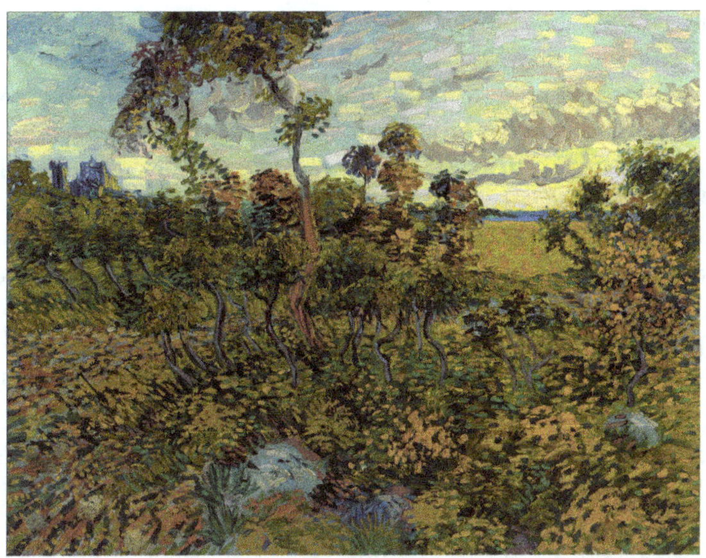

Fig. 31 *Sunset at Montmajour*

The scene van Gogh describes[4] in one of his letters is set in the forested landscape of Montmajour, in Provence, which he had often visited, attracted by "the view over the plain."[4] It was during his time at Arles in 1888, in the South of France, that van Gogh went on to immortalize its hilly landscape with a painting punctuated by his characteristic heavyset, broad impressionist brushstrokes and his lively coloured palette.

It promised to be another *chef d'oeuvre*, for had he not produced during such a period three of his masterpieces: *The Yellow House*, *The Sunflowers* and *The Bedroom*?

Yet, when he finished creating this work on a 93.3 cm × 73.3 cm canvas, he was not entirely satisfied with the results and confided to his brother that he believed it was "well below what I wished to do."[5] Van Gogh tended to destroy paintings he felt were below par, fortunately he spared this work (Fig. 31).

Until 1901, the work was owned by van Gogh's brother Theo. His widow Johanna then sold it to a Paris dealer. It was subsequently bought in 1908 by Norwegian industrialist Christian Mustad on the advice of art historian Jens Thiis.[6] Utterly bereft when a French Ambassador to Sweden told him he thought it was a forgery, Mustad banished the work to his attic. Almost 20 years after his death in 1970, the painting was sold in 1991 to a collector who contacted the Van Gogh Museum but failed to have it authenticated.

It would take another twenty years for its new owner (*name not disclosed as a matter of privacy*) to present it to the Van Gogh Museum in 2011, and this time the work was looked at with a different eye. The newly published compendium of van Gogh's correspondence was key; it revealed his mentioning of this painting in two letters to his brother Theo. In addition, a number 180 on the back of the canvas agreed with a number 180 on a list of van Gogh's oeuvre compiled after Theo's death by his brother in law. The number corresponded to a painting entitled *Soleil couchant à Arles* or *Sunset at Arles*.[1,6,7]

To validate the authenticity of the painting, the Van Gogh Museum made a stylistic analysis of the artwork and found many parallels with other paintings by the artist. First, they had a group of their own researchers analyse the pigments with a handheld portable XRF spectrophotometer[3] (page 150). Then samples were also prepared in cross-section and examined with a light microscope and by scanning electron microscopy and energy dispersive X-ray analysis (SEM-EDXRF) (page 138) confirming van Gogh's palette and technique.[3]

Furthermore, the newly developed computer algorithms (well-defined procedures that enable a computer to solve a problem) for weave canvas analysis (page 155) indicated that the distribution of both the warp and weft directions in the canvas agreed closely with another van Gogh painting *The Rocks* hanging at the Museum of Fine Arts, Houston (MFAH).

The Dutch experts who analysed the work, Louis van Tilborgh and Teio Meedendorp from the Van Gogh Museum in Amsterdam, called the discovery absolutely sensational in an article in *The Burlington Magazine*,[3] and Alex Ruger, director of the museum, exclaimed that it was "a rarity"[5] that a new painting could be added to van Gogh's oeuvre[3], "a once in a lifetime experience."[8]

In a press release it put out in September 2013, the museum announced:

All research indicates this painting is by van Gogh [...] *following two years of research into its style, technique, paint, canvas, subject, van Gogh's letters and its provenance.*

Rembrandt van Rijn (1606–1669)

The Telling Power of a Signature[1–5]

A painting — long thought to have been painted by one of Rembrandt's former students — was acquired in 2010 by the UK National Trust as a gift from the estate of Lady Samuel of Wynch Cross. With no place to hang it, it was taken to the former home of Sir Francis Drake, Buckland Abbey in Devon, where it was kept in storage for 18 months until its eventual display on the Abbey's dining room wall.

The 1635 painting is a self-portrait showing Rembrandt at the age of 29 looking at the viewer and wearing a dark black cloak and a cap with large white ostrich feathers. There is a metal band, presumably from a suit of armour, around his neck.

Initially, Rembrandt specialist Horst Gerso had doubted, back in 1968, that it was painted by the great master as, uncharacteristically, certain areas had been insufficiently worked on. In addition, the signature on the painting as well as the date of 1635 were not thought to be representative of the work he produced at that time or that age.

As other experts thought the self-portrait to be genuine, for decades its attribution was a hot topic. Chairman of the Rembrandt Research Project, Ernst van der Wetering — who had expressed doubt about the painting's authenticity almost five decades earlier — suggested in 2005 that it could, after all, be authentic.

In 2013, the painting was entrusted to the Cambridge Hamilton Kerr Institute, where it underwent eight months of close investigative work. It was carefully cleaned, with the removal of several layers of yellowed varnish. Once its original colours were revealed, it was subjected to a number of tests. These indicated that its pigments were consistent with Rembrandt's palette and that the wood of the panel belonged to the poplar/willow family in keeping with the kind used by Rembrandt. Additional tests carried out on the painting such as raking light photography, infrared reflectography and X-ray radiography (pages 133, 154, 151) pointed to an authentic work of art.

The most telling result was that obtained by a cross-section analysis (page 136) of the painting. A close scrutiny of the signature confirmed that it was not added after the execution of the painting, as described by the paintings conservator of the Hamilton Kerr Institute, Christine Slottvedd Kimbriel:

> *It was a close investigation of the artist's signature that gave us one of the biggest clues as to its true authenticity. The signature and date of 1635, inscribed both on the front and back of the panel, had been considered problematic in previous assessments as it was thought that the style and composition was much more akin to the artist's style slightly later in his career. But, the cross-section analysis left no reason to doubt that the inscription was added at the time of execution of the painting.*[5]

These results led van de Wetering to exclaim, "With all this additional scientific evidence, I am satisfied it is by Rembrandt."[6]

Johannes Vermeer (1632–1675)

Saint Praxedis: *A Surprise*

An image of contemplative serenity and pensive meditation of Saint Praxedis moves us deeply (Fig. 32). This powerful picture of an obscure second century saint, held in high esteem because of her protection of the bodies of Christians who perished under religious oppression, depicts a woman squeezing blood into a vessel from a drenched cloth apparently used on a beheaded martyr lying on the ground beside her.

This painting was displayed at the Metropolitan Museum of Art in 1969 as the work of the Florentine artist Felice Ficherelli (1605–1660), in an exhibit entitled *Florentine Baroque Art from American Collections*. The creation was thought to be another version of a similar 1640 painting by Ficherelli.

A signature on the left-hand side of the painting reading *Meer-1655* attracted the attention of art historian Michael Kitson who, having compared it to the signature on one of Vermeer's earlier works

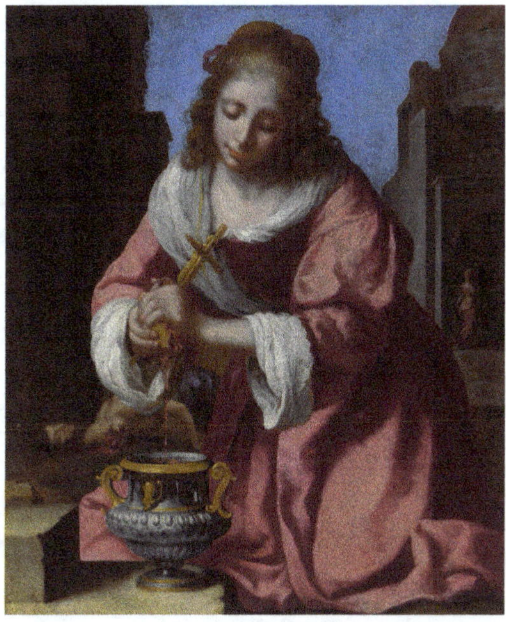

Fig. 32 *Saint Praxedis* by Johannes Vermeer

entitled *A Maid Asleep* attributed it to the Dutch master. Such an attribution astonished many, in view of its unusual subject matter in the wider Dutch context and of its atypical religious character given Vermeer's predominantly secular oeuvre.[1,2]

In 1986, Arthur Wheelock Jr (curator of the Northern European Collection at the National Gallery of Art in Washington) supported the attribution, having identified some stylistic analogies between *Saint Praxedis* and Vermeer's *Christ in the House of Martha and Mary* and some facial similarities with *A Maid Asleep*. Wheelock contended that at the young age of 22 or 23, the artist had copied a similar composition by Ficherelli when he had not yet fully developed his own very characteristic style, so admired today.[3]

In 1987, art collector Barbara Piasecka Johnson purchased the painting in spite of the scepticism of many experts with regard to the attribution. Connoisseurs suggested that the work had been created by a student of Ficherelli and that the added *Meer-1655* signature could not have belonged to Vermeer (van der Meer).

More than 10 years later, in 1998, the painting was still generally thought of as a work created in Italy either by Ficherelli or by some other Italian artist, and when an exhibition entitled *The Young Vermeer* was held in Edinburgh, Dresden and the Hague in 2010–2011, *Saint Praxedis* was not put on display. The painting was, however, included one year later in an exhibition of Vermeer's work held in Rome.

In early 2014, fresh analyses were carried out on the signature, the paint materials and on the lead white samples taken from the artwork. After careful scrutiny in a number of different labs, it was unequivocally confirmed that the signature was an integral part of the work and that the materials were in keeping with those used in Dutch paintings at the time. To confirm the attribution however, verification was needed that the work was created in Holland and not in Italy.

The recent technique of lead isotope ratio determination (page 146) was applied and gave compelling evidence supporting the attribution to Vermeer. Analysis carried out at the Rijksmuseum in association with the Free University in Amsterdam focused on minute amounts of lead white samples taken from both *Saint Praxedis* and from another contemporary and uncontested work by Vermeer, *Diana and her Companions*. The results indicated that the lead used in *Saint Praxedis* was of Flemish/Dutch origin and the paint so closely matched that used in *Diana and her Companions* that it was as if the same tube of painting had been used in both cases.[1]

Christie's, commissioned to sell the artwork and very pleased with the attribution, quoted Jørgen Wadum (chief conservator of the Royal Cabinet of Paintings Mauritshuis in The Hague) in its catalogue entry for the sale:

> *In the 2012/2013 Rome exhibition, Saint Praxedis was hung alongside the picture by Ficherelli which is now widely considered to be the (prototype) for it [...] The comparison perhaps posed more questions than it answered, not least as to whether another version or a copy of the Ficherelli might have served as the actual model for the Johnson picture. Rather than endorse the primacy of the Ficherelli, notwithstanding its somewhat abraded state, the comparison rather emphasized the expressive power and intensity of the Johnson painting. Indeed the exercise underlined one of the most disconcerting aspects about the Johnson painting in that this does not have the character of a formulaic replica*[1].

To Christie's delight, on July 8, 2014, Vermeer's *Saint Praxedis* was purchased by an unknown collector for $10.7 million.[4]

Leonardo da Vinci (1452–1519)

Salvator Mundi: *Recently Authenticated Treasure*

An extraordinarily beautiful, half-length frontal image of Christ (Fig. 33), his right-hand raised in blessing and his left holding a crystal orb, was identified in June 2011 as the original *Salvator Mundi* (*Christ as Saviour of the World*). In the time since, it has undergone meticulous conservation treatment, extensive technical analyses and scrutiny by prominent Leonardo experts. All results concurred: it was indeed a Leonardo masterpiece, which would make a valuable addition to the hitherto 14 known Leonardo oil paintings.[1–3] Infrared reflectography (page 154) was to play an important role in its authentication.[4]

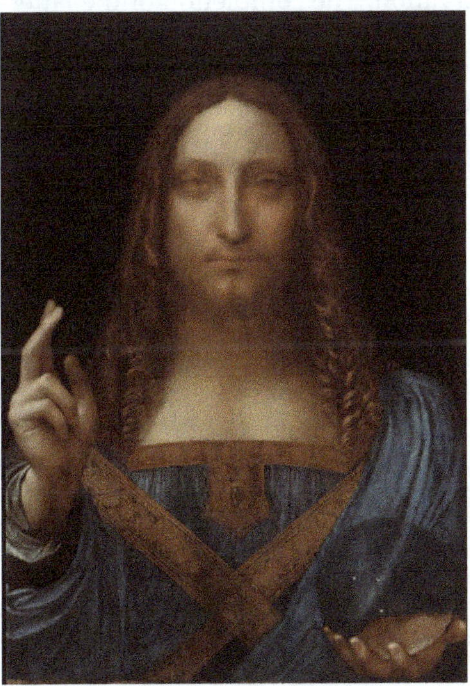

Fig. 33 *Salvator Mundi*

History of its provenance goes back to 1649 when it was first recorded as part of the art collection of King Charles I of England. The painting was sold at the latter's death and returned to the crown upon the accession of Charles II to the throne. It went on to become part of the Duke of Buckingham's collection and to be sold by his son in a 1763 auction following the sale of Buckingham House, later known as Buckingham Palace.[5]

There would be no further trace of the striking *Salvator Mundi* for over a century until its reappearance in 1900 (poorly blemished and disfigured) as the property of a Sir Frederick Cook. In 1958, the portrait was sold at an auction by Cook's descendants and catalogued as a copy of a painting by one of Leonardo's gifted students.[4] It remained as part of an American collection and, in 2005, was brought to private art dealer and art–historian Robert Simon for analysis and research.[5]

Renowned conservator Dianna Modestini was entrusted with the painting's conservation. The completion of the latter in 2007 revealed a work of astonishingly high quality by far surpassing any of the known copies of Leonardo's *Salvator Mundi.*

A key observation during the conservation process was the uncovering of *pentimenti* indicating that Christ's thumb had a more upright position than in the completed artwork. When the painting was then displayed to a large number of connoisseurs and art–historians, they all concurred that it did indeed display Leonardo's artistic genius. A significant and important seal of approval was given by prominent Leonardo specialist Martin Kemp who supported this attribution on stylistic, art historical and technical grounds.[4]

Kemp underscored the role played by infrared reflectography:

> *Examination by infrared reflectography, which involves bouncing infrared light off the white priming of the panel, has revealed characteristic signs of Leonardo's idiosyncratic technique. Particularly significant are clear signs of the palm of his hand into the wet paint, as can be seen above Christ's left eye. This was one of the techniques Leonardo used to create the soft, elusive effects for which he was renowned.*[4]

Had it not been for Modestini's conservation work[6] and the conclusive evidence it allowed experts to draw, *Salvator Mundi*'s recent astounding sale at Christie's in New York would have been most improbable.

The painting, referred to as the Holy Grail of Old Masters by a senior specialist at Christie's, had queues of people surrounding the Rockefeller Center just to look at the canvas. Many in the art world wondered if the piece would ever sell.

On 15 November 2017, with six bidders in play and rapid $20 million and $30 million jumps in price, an unusual sale seemed to be unfolding. After an initial pause at $200 million, a brisk bidding resumed with an offer of $332 million, which rapidly escalated to a final $450.3 million. The 20 minutes of dueling was followed by applause from people in the auction house cheering this historic purchase by a winning bidder later identified as a Saudi Prince from a remote branch of the royal family.

The news of the sale broke the world record for any painting sold at auction and was received with awe by many. To others, however, the sale expressed "another manifestation of a mad world in which nothing is madder than the machinations of the art market."[7]

Part II

The Court

Elena Basner: The Expert Under Siege

I have become the victim of some unpleasant, dark business which I myself do not understand.

Elena Basner[1]

Elena Basner's new scientific patented radiocarbon method (page 147) would threaten the murky market of art forgeries, embarrass art dealers, put an end to a very profitable business and bring "to light all the forged paintings now hanging on the walls of presidents, ministers, billionaire businessmen and multinational corporations."[1]

Could Basner (Fig. 34) have been the victim of a fabricated plot that led to her being arrested in January 2014 and charged with fraud and duplicity? Did unidentified enemies wish to see her disappear behind prison bars and shield her access to an emerging panoply of forgeries?

Her detention shocked the Russian art community: news of her arrest was followed by 2,400 people signing an online petition demanding her immediate release from prison in tandem with the exclamation of the Hermitage Museum Director, Mikhail Piotrovsky, that her placement in custody was an insult to "the whole Russian intelligentsia."[2–4]

All followed the unfolding drama with anguish and disbelief. Was the state prosecutor really serious when, two years following her arrest, he accused her of fraud in the sale of paintings and recommended that she be sentenced to four years in prison with a fine of

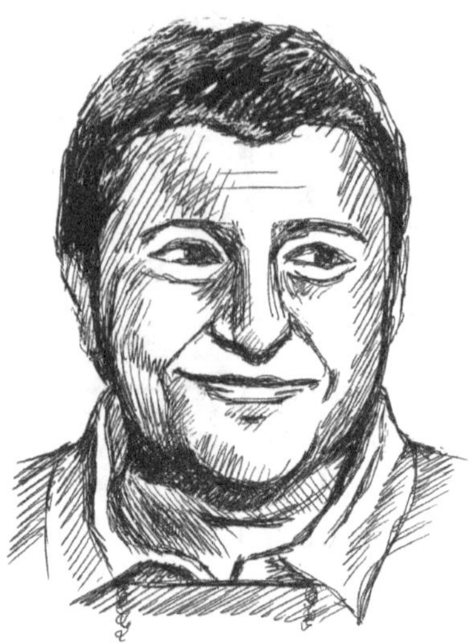

Fig. 34 Elena Basner (b. 1956–)

500,000 rubles?[5] How could this happen to Elena Basner, the acclaimed expert of international repute?

The Facts

It all started in May 2009 when Basner, a Russian avant-garde expert, was approached by a certain Mikhail Aranson who asked her to examine a gouache entitled *In a Restaurant* (Fig. 35). The painting was attributed to Boris Grigoriev, an artist who made a name for himself in the first half of the 20th century, as part of the Russian avant-garde movement aimed at freeing the imagination of both artist and viewer, permeated all of the arts, and in painting featured a variety of styles ranging from cubism to futurism to constructivism.

It looked like a pretty genuine version of the artwork, and Basner expressed the opinion that it did indeed seem authentic. Yulia Solonovich, another of the Russian Museum's experts, ratified Basner's opinion.

What Basner was unaware of is that a few months earlier, the Grabar Restoration Center had been commissioned to analyse that painting by an unidentified individual and had ruled it as a fake through the scientific detection of anachronistic phthalocyanine pigments using either Raman spectroscopy or FTIR (pages 139, 141).

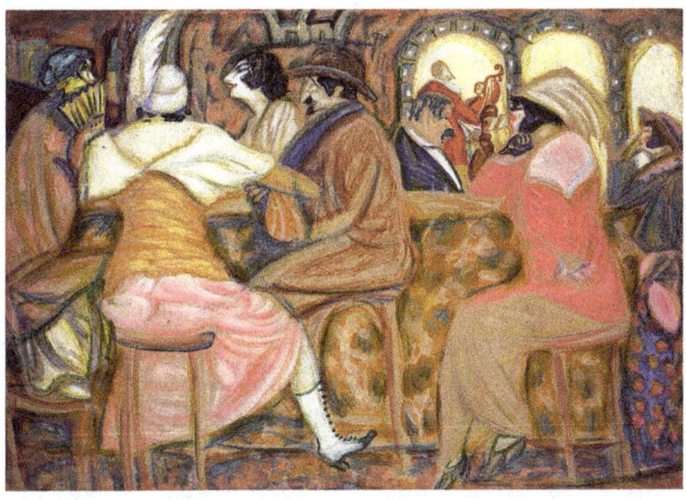

Fig. 35 *Paris café* by Boris Gregoriev

It is unclear if Aranson was aware of the verdict *inauthentic* on the painting by the Grabar Restoration Center and was planning Basner's downfall or whether it was just a mere coincidence that the painting was deemed a forgery shortly before he contacted Elena Basner. What is certain is that thanks to Basner's validation of *In a Restaurant*, Aranson could sell it for $180,000 to Leonid Shumakov, general director of the Golden Age publishing house, who in turn sold it to the experienced art collector and his long-time acquaintance Andrei Vasilyev for $250,000.[6]

Andrei Vasilyev's copy or version of Boris Grigoriev's *In a Restaurant*

Delighted with his new acquisition, Vasilyev lent it to an exhibition at the Kournikova Moscow Gallery to be admired by all onlookers. Little did he know that it would attract the attention of Yulia Rybacova who had closely analysed it earlier and branded it a forgery. The news would soon reach him...

Upon the realisation that *In a Restaurant* was a forgery and having found out that the original version of the painting entitled *Parisian Café* was kept in a storeroom in the Russian Museum in St Petersburg,[1,2] Vasilyev filed a civil suit against Shumakov in the Vyborg District

Court of St Petersburg and asked for the annulment of their contract and the restitution of the $250,000 he had paid for the painting. Alas, the suit was dismissed on the grounds that two years had passed since the initial purchase of the artwork as dictated by the statute of limitations.[1]

Vasilyev would not give up. He embarked on a rigorous tracking of the painting's recent history that finally led him to Elena Basner, whom he believed was the main perpetrator of the forgery. He accused her of fraud on a large scale and decided to initiate a criminal case.

It took several unsuccessful attempts before he finally succeeded in opening the case which would lead to Basner's arrest on 31 January 2014, and to her transfer to the Central District Remand Prison. She would be released from the detention centre four days later but kept under house arrest until January 30, 2015, when she was discharged with the provision that she would not leave town.

A Trial Rare in Its Dramatic Intensity

Faced with such accusations, Basner welcomed an inquiry; "I know I am innocent!"[2] she exclaimed, a thorough investigation was needed to clear her name and save her reputation.

The trial was bound to bring to light the following facts:

(a) A whole series of *Parisian Cafés* created in 1913 by Boris Grigoriev belonged to the Timofeyev Collection. General Nikolai Timofeyev had acquired them from art patron and collector Alexander Burtsey (1869–1938).[1]
(b) A special version of *Parisian Café* was bequeathed to the Russian Museum in 1984 by Kira Okunev who had inherited it from her father Boris Okunev an avid art collector of Russian art.
(c) The fact that Shumakov indicated to Vasilyev that the gouache came from a known local collection suggests that Elena Basner, who in the mid-1980s had frequently visited the Timofeyev collection, would have assumed that *In a Restaurant* was also part of that collection.[1] Basner had probably related that information to Mikhail Aranson, who had initially sought her opinion, and who in turn presented it to his client Shumakov as a proof of provenance.

(d) In making her evaluation of the gouache, it would be natural to assume that in 2009 Basner would have forgotten that *Parisian Café* was brought in 1984 (25 years earlier) and stored in the Russian Museum.

(e) Vasilyev's argument that *Parisian Café* had been stored in the depots of the Russian Museum since 1984, a time during which Basner was working at the museum, therefore accessible for her to have it copied, does not hold since:

  (i) Basner belonged to the department that dealt with oil or tempera paintings and not to the department that dealt with graphic art such as gouaches, and therefore she could not have had access to *Parisian Café*.[1]

  (ii) The Russian Museum strictly controls the procedure granting permission to artists wishing to make copies of paintings and drawings and keeps artworks on paper in locked drawers. Their removal from their storage place is meticulously documented and reported and known to all departmental staff members.[1,2]

(f) It is also noteworthy that until 2011, the only known available reproduction of Grigoriev's *Parisian Café* was in black and white, and to produce a convincing coloured copy one would have had to have access to the original.[1]

(g) In light of the above and of the Russian Museum's restrictions with regard to copying paintings in its collection, the most plausible explanation is that *Parisian Café* must have been copied or photographed in colour prior to 1984 when it was still privately owned by the daughter of collector Boris Okunev. Soviet collectors at such time displayed and sometimes loaned temporarily their artworks to other collectors[1] so that a coloured photograph of the gouache could easily have been made during such exchanges. This is an acceptable scenario in view of the following:

  (i) Coloured photography, invented by the 1908 Nobel Laureate Gabriel Lippmann, had become very popular and affordable in Russia and in the West as early as the 1920s.[7]

  (ii) The Vasilyev gouache *In a Restaurant* is distinctly different in its graphic details, size and composition from *Parisian Café*, and therefore could not have been a direct copy of the work in the Russian Museum[1] but possibly a version of it from a coloured photograph, which would account for the observed differences.

(iii) A good number of high-quality fake modernist artworks had permeated the Soviet art market in the 1980s.[8]

The Verdict

The Dzerzhinsky district court of St Petersburg began deliberations on Basner's case on May 11, 2016. To everyone's great relief, it did not take more than a few days after the start of the trial for the judge to acquit Basner of the accusations brought against her.[9–11] The announcement, acquittal, was received by ringing applause in the court hall with cries of "Thank God" and "Hurray, we won!"[1]

The Plight of Experts and the Russian Avant-Garde Market

As an expert on Russian avant-garde paintings of the 20th Century, Basner gave a totally informal opinion in her home on the gouache by Boris Grigoriev. She never expected that her personal attribution would lead to an avalanche of events culminating in a court case and her arrest.

Her plight justifies the current reluctance of experts and of reputable galleries to provide opinions related to the attribution of art works and illustrates the difficulties in dealing with the Russian avant-garde market.

Yekaterina Kartseva, co-owner and founder of Privatecollections.ru, expressed her frustration with the current impasse:

> *Less high-profile cases occur quite regularly; many of my friends who bought a painting at auction cannot obtain confirmation of authenticity at the Tretyakov Gallery. Major Russian auction houses are trying to keep track of their reputation, but many are bombarded with lawsuits.*[12]

The root of the problem lies in Russia's turbulent entry into the 20th century, the 1917 Russian revolution and the rise of the communist Soviet Union. The disarray that characterised that period saw the disappearance of many avant-garde paintings, with great Russian artists doomed to oblivion. The 1980s Perestroika and the ensuing fall of communism created a new elite conscious of the need to retrieve their own national heritage, together with the proliferation of billionaire collectors, which led to a very sharp rise in the price of these artworks.

Such an interest in avant-garde art, fuelled the black market and encouraged greedy forgers to produce a large number of forgeries.[8] As related in an interesting article in the highly regarded *Artnews* dealing with these present day Russian fakes:

> *With invented provenance, unreliable certificates of authenticity and 'rediscovered' works by artists who are lost to history, forgers are flooding galleries and auction houses with Russian avant-garde fakes.*[8]

To compound the problem, many forgeries were accompanied by certificates of authenticity issued by Russian institutions and art history experts, and works by minor Western artists were doctored to look Russian and questionable artworks were validated by being included in academic art books and sometimes reproduced in new catalogues. Some of the forgeries came from unknown private collections, whereas others were claimed to have been smuggled to Israel by immigrants in the 1970s. The bulk of the fakes were sold in Moscow, but an extremely large number of forgeries were also sold at sky-high prices in international galleries.

Art critic Simon Hewitt brings in another harrowing dimension to the problem:

> *The avant-garde is a case apart because the KGB set up something of a forging industry in the field as a way of accessing gullible foreign cash. They had access to materials dating from the 1920s which enabled many of their fakes to pass chemical testing with flying colours.*[12]

It is therefore no wonder that such a situation not only intimidates experts on Russian avant-garde art and makes them wary of giving their candid opinions, but it also jeopardises the natural workings of the art market.

Caravaggio Case: Plight of the Auction House

Because of the stakes involved, there remains an inherent legal risk to any act of authentication — no matter the precedent.

Mostapha Heddaya[1]

He looked with satisfaction at his recently acquired painting, it had been carefully cleaned and submitted to infrared reflectographic analysis (page 154), with delight he exclaimed to himself: *Yes it is a Caravaggio dating back to 1595, it is not as the auction house claims it to be a 17th century copy by a follower of the master, but an authentic Caravaggio*!

Collector Sir Denis Mahon looked again with awe at this beautiful painting entitled *The Cardsharps* which depicts an innocent-looking young man at a card table duped by two cheats. An older man has a split glove, which allows him to easily feel the marked cards and looking over the boy's shoulder, at his hand, he sends signals to his accomplice who holds a winning card tucked behind his back (Fig. 36); it looks very much like another Caravaggio, *The Kimbell Cardsharps*, worth £50 million, which is part of the collection of the Kimbell Art Museum in Fort Worth, Texas.[2]

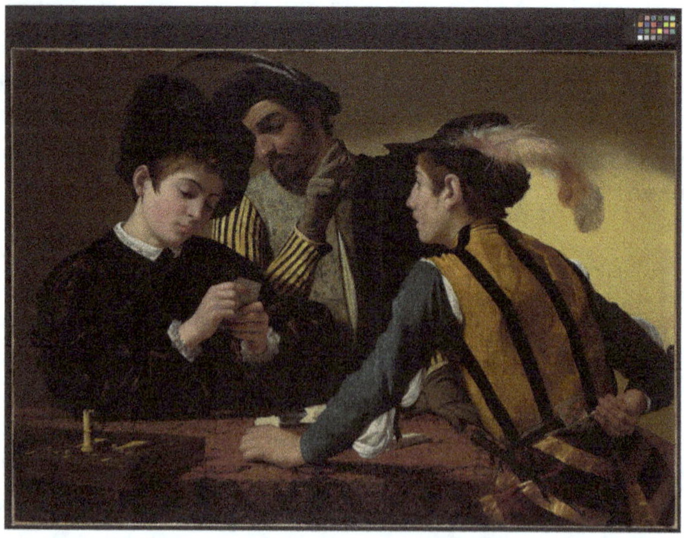

Fig. 36 *The Cardsharps*

Thrilled by his discovery, Sir Denis proudly announced to the world in 2007 (on his 97th birthday) that the painting bought on his behalf one year earlier by Mrs Orietta Adam from Sotheby's for only £42,000 was really worth £10 million. According to him, it was an autograph replica painted intentionally by Caravaggio to be identical to the original version.[3]

He could not have felt more certain of the matter. Had he not, together with the highly regarded art historian Professor Mina Gregori, authenticated *The Kimbell Cardsharps* upon its rediscovery in 1987 at the Zurich Institut für Kunstwissenschaft following its disappearance in 1890?[4] Had the art establishment not accepted his past authentication of two Caravaggios in the 1970s: *Martha and Mary Magdalene* and *The Crucifixion of Saint Andrew?*[5] No connoisseur in the world would dare challenge his attribution!

Sir Denis's announcement came as a shock and surprise to Mr Lancelot Thwaytes who had inherited Mahon's version of *The Cardsharps* in 1965 from his first cousin William Glossop Thwaytes, a navy veteran and Caravaggio collector who had bought it in 1962 for only £140. Aware that his cousin had previously purchased *The Musicians*, which was subsequently identified in 1951 by Mahon and Gregori as a genuine Caravaggio, Lancelot suspected that his version of *The Cardsharps* was also authentic.

In 2006, he had commissioned Sotheby's to assess the artwork's authenticity. Tom Baring, the junior specialist at Sotheby's who had been tasked with its evaluation, concluded that the painting was a good 17th-century copy of *The Kimbell Cardsharps*. Baring had also shared his conclusions with two senior Sotheby's experts[6] who concurred with him that it was indeed a copy.

When informed of Sotheby's evaluation, Mr Thwaytes requested that further X-ray *and* IR reflectographic investigations be made. Sotheby's carried out the X-ray tests, but instead of the infrared, UVF (page 239) was applied as was standard auction practice. *(X-ray radiography and Infrared reflectography penetrate the surface of the paintings to analyse the so-called underpaintings which are the initial stages of the creation of the artwork. In general, infrared reflectography gives much more detailed information on these underdrawings, indicating whether or not the artist changed his or her mind in terms of the positioning of a given figure; these changes are referred to as 'pentimenti'. UVF merely allows for the surface examination of the painting, indicating areas of retouching.*)

As no new evidence supporting authenticity was provided by these tests, Sotheby's informed Mr Thwaytes that it was maintaining its initial opinion, the work was a copy and there was no need for further investigation.[4]

There seems to have been a certain amount of confusion between Mr Thwaytes' request and what was intended to be carried out, as Sotheby's team confirmed that the painting would undergo UVF and X-ray tests, which seems to have been understood by Mr Thwaytes as infrared and X-ray analyses.[4]

At that point, Mr Thwaytes decided to sell the painting, and it was mutually decided that it would be placed in an Old Masters Painting auction sale taking place in London on December 2006, in Sotheby's Olympia auction room and described as a 17th century copy after the Kimbell original *Cardsharps*.

What was not related to Mr Thwaytes was that shortly prior to the auction, at a sale exhibition in the Olympia show room where potential buyers interested in Old Masters could inspect the paintings and talk them over with the experts, the level of interest the painting had attracted was tremendous, which prompted Mr Baring to seek a further opinion from Sotheby's experts at New Bond Street in a special meeting in Olympia.

These in turn unanimously concurred that the painting was indeed a copy and not an autograph replica. The initial pre-sale value was estimated at £20,000–£30,000, and the painting was eventually sold for £42,000.[4]

Justifiably upset by Sir Denis's announcement, Lancelot Thwaytes sued Sotheby's on the grounds of negligence and breach of contract. He was dismayed by the fact that the auction house had reached the conclusion that the work was not by Caravaggio without carrying out the tests that he had required them to do, in particular failing to use IR imaging. He also criticised Sotheby's for not informing him about an unusual interest in the painting, which led to its re-examination by additional Sotheby experts. He argued that he might have decided not to sell the painting at all or he might have sold it at a higher price.

He supported his arguments by pointing out additional features in the construction of the work including some observed *pentimenti* (page 227) as revealed later by IR imaging, which indicated that it was not a copy.

When testifying in high court he exclaimed:

Words cannot really do my emotions justice but I was in utter disbelief and absolutely horrified to see that the painting was now being proclaimed to the world as an original Caravaggio, little more than a year after the auction. I thought by asking Sotheby's to properly research the painting, and by asking them repeatedly if they were sure that it was a copy, that I had done everything that I could in my position. I felt extremely let down and very angry that Sotheby's had apparently not done their job properly.[7,8]

The proceedings of the trial which started in October 2014 in the English High Court before Mrs Justice (Vivien) Rose, with representatives of both parties resting their case, lasted 17 days.

The legal team for Mr Thwaytes clarified from the start that the crucial parameter in this trial was a claim of negligence, that Sotheby's had not carried out a thorough investigation of the artwork and had not sought the opinion of external experts as it should have. Proving the authenticity of the painting was not the main issue at hand in this case.

After listening carefully to the arguments on both sides, Mrs Justice Rose concluded that Sotheby's was not negligent and justified her final ruling in a brilliant exposé where she clarified why she had reached such a verdict. Unlike other judges who are generally reluctant to reach a decision with regard to questions of art authenticity, she discussed her own evaluation and analysis criteria with regard the authenticity of *The Cardsharps* and wrote:

I bear in mind Buckley J's warning in Drake v Thos. Agnew & Sons Ltd [2002] *EWHC 294 (QB) about substituting my own assessment of quality for that of the experts. However, it seems to me that the task is inescapable here, given the issues in this case.*[8]

The following salient points explain why Sotheby's specialists, in her view, had carried out and fulfilled their duty of care with due diligence and in a satisfactory manner. In her opinion, Sotheby's, as a renowned auction house, did not have any obligation to consult outside experts and was entitled to trust the evaluation of its own highly qualified art consultants:

> *I consider that those who consign their works to a leading auction house can expect that the painting will be assessed by highly qualified people — qualified in terms of their knowledge of art history, their familiarity with the styles and oeuvres of different artists, and in terms of their connoisseur's 'eye'.*[8]

As for the fact that the auction house did not inform Mr Thwaytes of the extra interest that the painting had aroused cannot be considered negligent, since the opinion of Sotheby's Bond street experts concurred with the earlier ruling. Such a conclusion would most certainly not have led to a change of mind on the part of Mr Thwaytes with regard to selling the painting.

The judge pointed out that the reading of the X-rays was done correctly, with no indication of any additional information that would support authenticity, and that the auction house was under no obligation to carry out IR analysis.

Through a clever and perceptive analysis, she went on to explain why the *pentimenti* which were detected by IR testing as requested by Sir Denis were not of a kind that would suggest authenticity and why they could have been made by any copyist.[8]

Moreover, the information provided by Sotheby's lawyers suggesting a change in Mahon's mental faculties at the ripe old age of 96 and questioning his ability in evaluating an artwork cast doubt on his authentication of *The Cardsharps*. Hadn't he and Dr Mina Gregori given an initial negative attribution with regard to *Saint John at the Well* attributed to Caravaggio, and then changed their mind once the painting was cleaned and analysed by infrared? Is it not telling that Sir Denis could not recall that the negative attribution ever happened?[8]

As for Dr Gregori's evaluation of the painting, Sotheby's experts did not place much weight on her expertise for Caravaggio's works,[8] previously and under different circumstances she had supported a painting bought by an art collector as a Caravaggio that turned out to be by Bartolomeo Cavarozzi.[8]

It would be unjustifiable to impose on Sotheby's the spending of resources as well as precious time researching artworks that had not yet been entrusted for sale, she emphatically stated in her judgement.

Following the announcement of the verdict, Mr Thwaytes expressed great dismay at the court ruling and utter astonishment that the court had rejected his claim that he had expected a higher standard of care from the auction house.

Relieved and delighted with the verdict, Sotheby's made the following announcement:

> *Sotheby's is delighted that today's ruling dismisses all claims brought against the company and confirms that Sotheby's expertise is of the highest standards.*[9]

Plight of the Auction House

There is no doubt that *authentication* is of crucial importance in the purchasing of an artwork and provides the seller with a certain degree of confidence and security when dealing with an auction house. However, as was the case with *The Cardsharps,* there is always the threat that having followed the advice and evaluation by the auction house, the seller would hear a different opinion on the work after its sale and pursue legal action.

Judges are generally reluctant to make decisions on authenticity matters and rely on the testimony of the experts representing both sides of the conflict. In *The Caravaggio Cardsharps* case, it is interesting that a claim of negligence and breach of contract should have instigated a deliberation on *authenticity* issues as discussed and put forward by Mrs Justice Rose who, with extraordinary insight, played the role of the connoisseur, art historian and expert at the same time. Such an exercise assisted her in her ruling of the case.

This is even more impressive as evaluating art is sometimes a daunting task, especially in the case of old masters, the likes of Caravaggio, since the most sophisticated scientific tests cannot always give a clear answer with regard to authenticity as the work could have been created by a student or follower of the alleged artist.[10]

Mrs Justice Rose closely looked at the painting, drew her own conclusions with regard to authenticity, carefully analysed the approach of the auction house in dealing with the case and finally reached her verdict. The latter addressed questions raised with regard to the extent of responsibility and duty of a reputable auction house when a painting is consigned to it.

She ends with the following very telling conclusion:

> *In my judgment there is no basis for concluding that when a work is consigned to an auction house for research and assessment rather than for sale, that imposes on the auction house a duty to examine the work more carefully than they need to if it is consigned to them for sale*[8]

Can *The Cardsharps* case assuage the fears of reputable auction houses while guiding them in terms of what is expected from them? Can they rest assured that their evaluation of an artwork will not carry legal risk, or will claims of negligence and the spectre of litigation always loom ahead?[11–14]

The Knoedler Case: Plight of the Collector

> *I got a fake painting for $8.3 million and I want my money back!*
>
> Domenico de Sole[1]

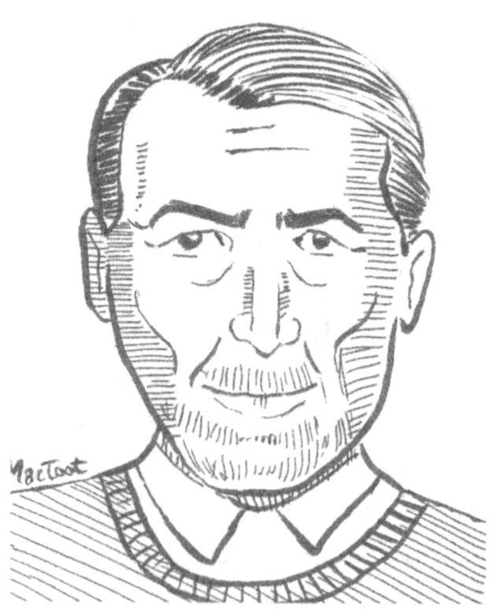

Fig. 37 Domenico de Sole (b. 1944–)

The Rothko Forgery

As a means of fairly dividing their estate between their two daughters, late in 2004, Domenico de Sole (Fig. 37) and his wife Eleanore decided to purchase a painting by the renowned Irish artist Sean Scully.

They had seen one of his pieces hanging on a friend's wall, found it very appealing and were keen to purchase another of this artist's paintings.

However, there were none available for sale at the highly reputable Manhattan Knoedler Gallery, and the de Soles were persuaded by its president Ann Freedman to buy instead a painting by the abstract expressionist painter Mark Rothko.[2] Its red and black palette was very striking, and it was attractive in its simplicity, meticulous attention to colour, composition and perfect balance (Fig. 38).

The de Soles bought it for $8.3 million, carefully encased it in an expensive glass equipped with an alarm system and shipped it to South Carolina where they hung it in their home. For many years, it would be a source of great pride for its owners whose numerous friends viewed the prized acquisition with awe.[3]

Fig. 38 *Untitled 1956*, forged Mark Rothko

All of this came to an abrupt end in late 2011. Domenico de Sole had just finished taking his shower one morning when he heard his wife scream, he rushed out only to find her shaking almost hysterically and crying. She had just read in *The New York Times* the alarming news that a painting *Untitled 1950* of questionable provenance and attributed to Jackson Pollock was a forgery. The Knoedler Gallery had sold it in 2007 to hedge-fund manager Pierre Lagrange for $17 million. The article indicated that the artwork was supposedly owned by the son of an anonymous Swiss collector who had purchased it through the help of the late gallery dealer and art–adviser David Herbert.[4] The alleged Pollock provenance of the painting was too familiar for comfort as it sounded very much like that of their precious Rothko.

De Sole wasted no time in hiring forensic scientist James Martin from Orion Analytical who closely analysed their painting. Micro-FTIR measurements (page 141) indicated that a water-based polyvinyl acetate was employed as a base which was never used by Rothko, and further investigation indicated a white opaque primer layer, not a Rothko technique at the time.[5]

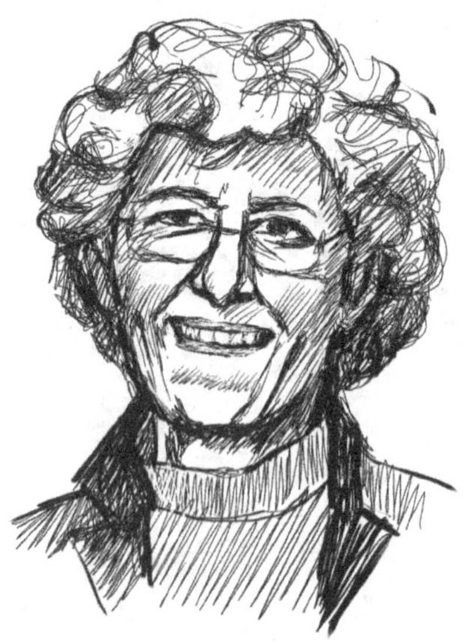

Fig. 39 Ann Freedman

They had been duped! Domenico and Eleonore de Sole did not hesitate to initiate a lawsuit accusing the Knoedler Gallery and its president Ann Freedman of fraud (Fig. 39).

Glafira Rosales

The story goes back to Glafira Rosales, a Long Island art dealer who filed false tax returns and deliberately failed to disclose an offshore bank account hiding at least $12.5 million dollars from sales of unknown Abstract Expressionist paintings attributed to famous painters such as Jackson Pollock, Mark Rothko and Willem de Koonig.[6,7]

Forty of such works, created by Pei-Shen Quian, a gifted Queens Street Chinese artist, were sold by Rosales to Knoedler & Company, which the Gallery then resold at astounding markups. These sales would help the Gallery overcome its temporary financial difficulties and bring it back into the black.

Rosales claimed that she represented the heir of an anonymous Swiss Collector — a certain "Mr X" — who lived in Mexico and Switzerland, and who had apparently bought the collection from the artists themselves through the help and advice of the Filipino American Abstract Expressionist artist Alfonso Ossario.

However, when in 2001 Jack Levy, co-chairman of mergers and acquisitions at Goldman Sachs, decided to buy a Jackson Pollock for $2 million from Knoedler and Company, his precondition for the purchase was that the work be examined by the International Foundation for Art Research (IFAR). Investigations by IFAR revealed that the story of the defunct Ossario was very dubious, the latter not alive to either ratify or deny it; furthermore the provenance was vehemently challenged by Ossario's long-time partner Ted Dragon. IFAR was also concerned about the signature, style of the painting and absence of documentation.[8]

The deal was therefore not carried out, and Levy recovered his $2 million. Rosales subsequently changed her story claiming that she had misunderstood, and that the real advisor who sold paintings to "Mr X" was the late dealer David Herbert who had worked at the Betty Parsons and Sidney Janis galleries in the 1950s.[9]

Rosales' actions would eventually lead to her arrest and, as creatively pointed out by United States Attorney Mr Preet Bharara, she gave a "new meaning to the phrase artful dodger by avoiding taxes on millions of dollars in income from dealing in fake artworks for fake clients."[10]

The Trial

Pierre Lagrange would first question the authenticity of his $17 million Jackson Pollock painting purchased from Knoedler Galleries in 2011, four years after he purchased it, finally leading to the revelation which would later arouse the de Soles' suspicions in relation to their Rothko. Lagrange became suspicious of his costly acquisition as a result of Sotheby's and Christie's refusal to sell his painting given that it was not included in Pollock's catalogue raisonné. To assuage his doubts, Lagrange hired forensic scientist James Martin to investigate the painting. Analysis of the work by FTIR revealed the anachronistic Pigment Yellow 74, not commercially available during Pollock's lifetime. Enraged, Lagrange contacted Freedman and gave her 48 hours to repay his money.

This led to the abrupt shutting down in 2011 of the one-time respected and trusted 165-year-old Knoedler & Co Gallery and was followed by a large number of forgery-related lawsuits against both the gallery owner and Ann Freedman. Lagrange's charges of fraud would be settled one year later in October 2012,[11] followed by other litigation settlements.

The only lawsuit that did go to trial was that of Domenico and Eleanore de Sole. This legal action relied on a line of experts and unleashed a mountain of embarrassing evidence and incriminating testimony describing how dozens of collectors, including the de Soles, were deceived into buying forged artworks attributed to Abstract Impressionist masters (Fig. 40).

James Martin who analysed a total of 16 of Rosales' paintings, found that the majority had anachronistic materials, some had dubious signatures and others showed signs of intentional ageing.[12] He exclaimed that, taken in toto, the anomalies were comparable to "JFK holding an iPhone!"[5]

Through probing cross-examination, the unfolding of the trial would reveal a string of red flags that Freedman wittingly or

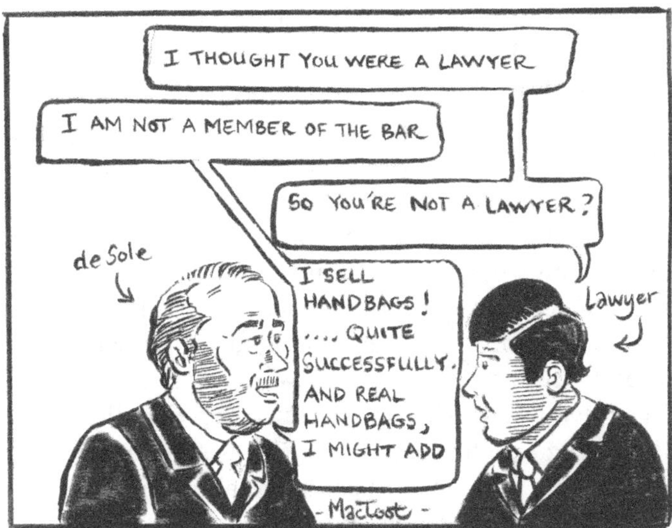

Fig. 40 Lawyer interrogating Domenico de Sole

unwittingly ignored as well as a series of eyebrow-raising actions on her part.

To everyone's regret, this riveting drama would come to an anticlimactic ending when, in a courtroom crammed with people brimming with curiosity and expectation, the judge abruptly stayed the proceedings. A settlement between plaintiffs and defendants had been reached out of court.[13]

Red Flags

The fact that Rosales offered Freedman artworks well below market prices should have aroused Freedman's suspicion from the outset. The Gallery bought the Rothko in question for $950,000,[2] a notably low price for such an artist. Could Freedman have been totally oblivious to this fact or had she decided to merely relish her success at selling it to the de Soles with such an astonishing markup?

The paintings surreptitiously flowed in over the years, one by one, from this Secret Santa as Freedman would come to refer to "Mr X".[14,15]

Such was her enthusiasm for this gold mine of artworks that she missed a wrong spelling of Pollock as *Pollok* and dismissed the concerns expressed by Gretchen Diebenkorn, daughter of renowned California artist, Richard Diebenkorn, with respect to two paintings attributed to her father. Gretchen shared her worry with Freedman about the absence of provenance of the works, which she described as "bland" and "with no soul."[16] To the astonishment of the Diebenkorn family, the paintings were subsequently sold.

Likewise, Freedman did not seem to worry that her researcher, Dr Melissa de Madeiros, found no evidence or documentation that David Herbert was acting as a consultant or sold artworks to a Swiss collector who lived in Mexico and Switzerland. She disregarded forensic tests indicating anachronistic pigments as she deemed them meaningless;[6] apparently, she was unpersuaded about IFAR's report on the purported Jackson Pollock and did not question Rosales' credibility when the latter changed the provenance from Ossorio to David Herbert. Furthermore, Freedman did not pay much attention to Rosales's unusual cash settlement request asking for payment to be made partly in cash *under* $10,000, partly by cheque and partly by wire transfer.[17]

At the trial, Martha Parrish, an old-time art dealer who participated in preparing the Art Dealers Association of America's (ADAA) code of ethics, testified that Freedman should have given much greater weight to all of these irregularities: "A reputable and responsible dealer would run like hell, because there are too many red flags flying for anyone to take the risk."[18]

A Closer Look and Analysis at Some of Freedman's Actions

In offering the Rothko for sale, Freedman referred to it as a masterpiece[2] and claimed in a letter to the de Soles that 11 experts had "viewed" the painting, with the implicit inference that it had received their seal of approval. Included in the list were experts on different Abstract Expressionist artists including Rothko and Motherwell. When testifying at the trial, several witnesses expressed astonishment at being included in the list of 11 experts, some claiming they did not even remember viewing the artwork.

When Lagrange rejected the Pollock, Freedman dismissed the forensic test results, claiming that in the 1950s artists were given trial paints long before they became commercially available; according to her these were "mystery paints that weren't registered that people used."[19] She suggested to an incensed Lagrange that she would find another buyer because in her view the painting was perfect.[9]

Freedman adamantly maintained that she believed the paintings were authentic when interviewed in April 2012, more than a year *before* Rosales confessed[20] that she had sold fakes.

"Because we haven't yet sorted out the provenance doesn't mean there's no provenance."[9] she confidently maintained. An indication that she believed in the authenticity of Rosales' artworks can be gleaned from the fact that she traded paintings from her own collection for some of the latter,[17] and bought three more for herself. Presumably, she convinced herself that if a dealer viewed the Rosales paintings and did not express any reservations, then it equated to approval.

Referring to top dealers who had viewed her paintings, she asserted, "Had anyone found anything wrong, meaning they weren't 'right,' … believe me, I would have been told 'Take that down off the wall'. No one ever did."[9] And if it happened that some incredulity was conveyed by a dealer, Freedman attributed it to professional jealousy![9]

In the same interview, she claimed with assurance:

> *Thank God that I didn't just let these paintings die because no one had the courage of their convictions. When all this comes out to be true, I'll be thanked. That's what I believe.*[9]

As irrational as this may seem in hindsight, a closer look at Freedman's behaviour suggests that her interpretation of all the evidence that came before her was influenced by her own belief in the authenticity of the works. She weighed information selectively, giving greater weight to facts that supported her belief and less weight to facts that cast doubt on the provenance of the paintings. Her behaviour is in keeping with what psychologists refer to as *confirmation bias* entailing:

> *The tendency to search for, interpret, favor and recall information in a way that confirms one's preexisting beliefs or hypotheses, while giving disproportionately less consideration to alternative possibilities.*[21,22]

Such bias was prophetically identified by Francis Bacon way back in 1620 when he wrote in his *Novum Organon:*

> *The human understanding when it has once adopted an opinion (either as being the received opinion or as being agreeable to itself) draws all things else to support and agree with it. And though there be a greater number and weight of instances to be found on the other side, yet these it either neglects and despises, or else by some distinction sets aside and rejects.*[23]

Freedman took advantage of the prevalent and accepted notion that in the art market, sellers and buyers are often anonymous as "if that gave her a license to believe everything that Glafira Rosales supposedly said."[17] She did not question the low price of the paintings, assuming as Rosales had explained that the owner came from a well-to-do family and just wanted to do away with the artworks.[24]

Freedman's statements during the 2012 interview, together with those she made in an exclusive 2016 interview[24] after the trial came to an end, appear to confirm that her actions were indeed fuelled by this type of cognitive bias: it entailed favouring any information that supported her existing belief and discrediting that which opposed it.

Talking about the red flags referred to during the trial, she claimed that these "gave me more fire to be proving better and harder what I believed."[24] Reinforcing her strong past conviction as to the authenticity of the works, she said, "I always thought that tomorrow I'll find a photograph of the artist's studio and see the corner of [one of the forged paintings.]"[24]

Reviewing the whole situation retrospectively, she admitted:

> *Looking back, there can be things I didn't see at the time* [...] *Could I have done some things differently? Not a day goes by that I don't think about it. I don't have an answer sitting here. I will at some point probably.*[24]

Plight of the Collector

The de Soles went into a prestigious 165 year-old gallery of impeccable reputation, were offered a painting by a renowned artist and were given as a warranty a list of 11 Rothko experts who had viewed the work with the implicit inference that they had also approved it. The painting was subsequently bought, without any further probing on the part of the de Soles who trusted Ann Freedman unwaveringly and relied on Knoedler's reputation.[25]

There is no doubt that, after the Knoedler scandal, collectors will be more vigilant and less trusting; however, the question remains as to whether or not in future they will need to overturn long-standing industry practice and carry out their own investigation regarding the authenticity of the artwork as well as validating the purported provenance for their own peace of mind. This might prove to be an expensive and daunting task, considering that experts are generally reluctant to express an opinion for fear of being sued for libel or slander by interested parties.

In the aftermath of this trial, what should have been done can be easily surmised. More problematic, however, is deciding on future solutions.

As judiciously put by Eileen Kinsella:

> *Never has the phrase "Hindsight is 20/20" held as much weight for anyone as it has for collectors and other art world observers in the wake of the recent high-profile Knoedler forgery trial.*[19]

Safeguarding the Art Market

If you look at the art market today, the lack of proper record keeping and the inability to run real-time verification of provenance have created a ripe environment for fraud.

Robert Norton[1]

Expert opinions on authenticity play a crucial role in safeguarding the art market from being flooded with forged works of art. Avid collectors need guarantees that they are purchasing the real thing, and newly discovered artworks need to be appraised. However, cases described earlier reflect some of the legal liabilities and high stakes involved in any act of authentication.

Such risks have led to the dissolution in 2012 of the Authentication Board of the Andy Warhol Foundation claiming that it had faced defence costs of more than $10 million for a series of suits in which its own viewpoint had been eventually endorsed. In a similar vein, art experts have frequently kept silent on authenticity matters for fear of libel.

The proliferation of lawsuits against authentication foundations, gallery dealers, authors of catalogues raisonnés, auction houses and experts who provide their candid opinions on artworks threaten not only the natural functioning of the art market but may also lead to its freezing and eventual downfall.

So, what can be done to reassure authenticators that they will not be subjected to prohibitively expensive, tortuous lawsuits, and how can collectors, gallery dealers and auction houses be reassured that they are dealing with *authentic* works?

Protecting Authenticators

In an attempt to protect art authenticators, the New York State Senate passed a bill on 15 June 2015, entitled: *An act to amend the arts and cultural affairs law, in relation to opinions concerning authenticity, attribution and authorship of works of fine art.*[2,3] The bill renders litigation against authenticators more difficult, and can considerably reduce the legal costs when authenticators are faced with a lawsuit (Fig. 41).

This bill defines an *Authenticator* as … *a person or entity recognized in the visual arts community as having expertise regarding the artist, work of fine art, or visual art multiple with respect to whom such person or entity renders an opinion as to the authenticity, attribution or authorship of a work of fine art or visual art multiple, or a person or entity recognized in the visual arts or scientific community as having expertise in uncovering facts that serve as a direct basis, in whole or in part, for an opinion as to the authenticity, attribution or authorship of a work of fine art or visual art multiple. Authenticator shall include, but not be limited to, authors of catalogues*

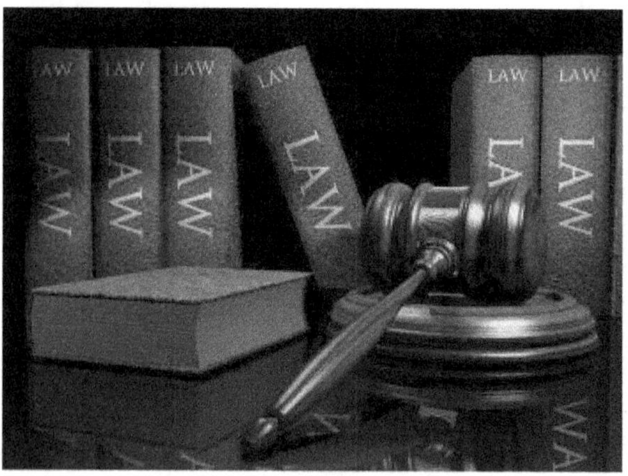

Fig. 41 The rule of law

raisonnes or other scholarly texts in which an opinion as to the authenticity, attribution or authorship of a work of fine art or visual art multiple is expressed or implied. The bill goes on by expressly excluding as an authenticator "a person or entity that has a financial interest in the work of fine art."[3]

The bill has to be approved by the New York Assembly and eventually by the New York Governor.[4] So far, the Senate voted overwhelmingly in favour of the bill when it was presented in the 2015 and 2016 sessions, but the committee responsible for presenting the bill to the Assembly floor for a vote has held it up. It is hoped that it will be reintroduced to the Senate and Assembly in 2017/2018 to be enacted into law.[2]

According to this bill, owners who give their opinion on an artwork or other experts that have a financial interest in the artwork or the transaction do not qualify as authenticators. The requirement that the authenticator should be a recognised expert either by the scientific or visual arts communities is not precise and leaves it to the courts to decide the qualification of a would-be authenticator where community recognition is questionable.

The bill allows an authenticator who prevails in a lawsuit to recover his or her legal fees and costs. Whether to award such fees and costs to a prevailing authenticator, however, is left to the discretion of the judge, and the authenticator would only be informed of reimbursable costs at the end of the case.[3,5] It is not clear therefore whether this part of the bill will provide authenticators with sufficient financial guarantees to incentivise them to give authentication opinions.[5]

It remains to be seen if, based on this Act, authenticators will feel less vulnerable, and sufficiently backed-up to resume their assessment activities.

Safeguarding the Art Market for Present and Future Generations[6–10]

A Synthetic Bioengineered Encrypted DNA-based Identifier

A recently developed industry standards-based, bioengineered DNA-based technique may provide an answer to all these uncertainties, safeguarding the art market and protecting authenticators. It consists of embedding into various forms of artwork (paintings, sculptures, etc.) a synthetic bioengineered encrypted DNA-based identifier produced in the lab, and linked to other informatics technologies, which would act as a permanent marker for the artist's work upon its creation or discovery. This kind of solution will enable creating an accurate history associated with the exact object when it is created, or at a defined point of intervention for already existing secondary market artworks, and link the object to the continuing information surrounding the object.

This *permanent and concealed marker* (initially costing around $150 and then expected to be significantly cheaper on mass production) would be imprinted on *labels* or *tags* (with encrypted digital equivalents for digital art, for example), which in turn would be affixed to the art object in a non-harmful (zero impact) way. For physical artworks, through materials science the marker would be transmitted to the latter, at the molecular level, where it would remain even if the label is withdrawn. The label enables verified users, including artists, through various other encryption and verification protocols, to interact with the artwork through an outwards-facing element of the label and through secure technology applications, which enable a cumulative verification of everything happening to the identified object as time goes on, in an Internet of Things (IoT)-like fashion (see Glossary). Importantly, this creates uniqueness based on the object and based only on the owner or creator.

The reason for using bioengineered as opposed to the artist's personal DNA is that this approach addresses privacy issues arising from holding records of the DNA profiles of artists and is far more secure than an individual artist's DNA, which can be stolen (in ways as simple as shaking the hand of an artist) and implanted in a forged artwork (Fig. 42).

This unique collection of technologies offering secure features and protocols along with verified and encrypted information was initiated at the Global Center for Innovation for the i2M Standards based at the State University of New York, Albany. It has led to the important i2M project, which has received generous funding ($2 million) by the ARIS Title Insurance Corporation exclusively dealing with art. The goal of the Center is to provide a standards-based framework known as i2M to support technology-based

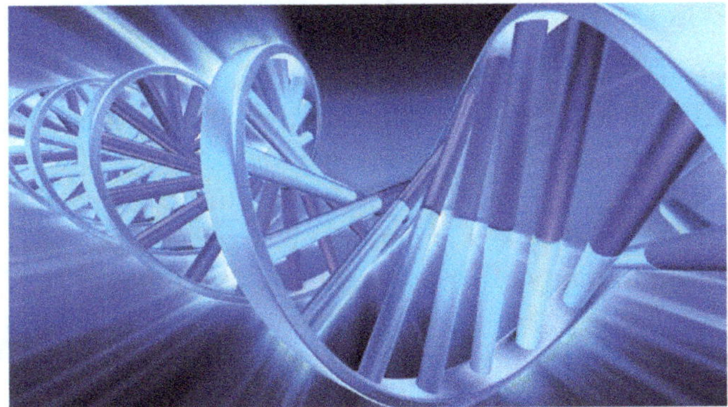

Fig. 42 DNA

"industry-wide solutions"[7] to the problem of forged art. According to the i2M Center it would be virtually impossible for skilled forgers to decipher, replicate or replace the engineered DNA-based marker and its associated verification elements.

As related by Lisa Karczewski:

> *According to the developers, the bioengineered DNA would be "unique to each item and provide an encrypted link between the art and a database that would hold the consensus of authoritative information about the work." A scanner, accessible to anyone in the art industry wishing to verify a work, could read the details of the DNA embedded in the artwork.*[6]

Such standards-based procedures would really be the Holy Grail for all artists and collectors by enabling irrefutable authentication or deauthentication of an art object. A number of renowned artists have already expressed great interest in using this new kind of approach.

Blockchain: A Revolutionary Technology

A blockchain acts as a public database or ledger that is cryptographically secure and on which data records and transactions have been permanently stamped and registered, instantly appearing on a vast network of computers.[11] Data records can neither be revised nor altered.

How It Works

To understand more clearly how a blockchain works, assume that I have one exclusive Cartier pen. I give it to you, then it belongs to you, and only you have absolute control over it.

Now assume that I have one digital Cartier pen and that I give it to you; this time, because it is digital, how can you be sure that I did not also give it to many other friends and that it is *only yours*? Maybe I made copies of this pen and put it on the internet where it is accessible for downloading by a large number of people. In such a situation, how can you possibly be certain who else has a copy of your digital pen?[12]

A Solution: Public Ledgers

If these digital pens are tracked in a kind of accounting book, a public ledger that is registered in a vast network of computers in the system, all the exchanges related to Cartier digital pens in this particular network will be precisely recorded. As the overall number of pens has been initially defined in this public ledger, any other additional pens would be out of sync with the system. The exchange of a digital pen therefore becomes as restricted as that of a physical pen, eliminating any uncertainty.

Now, if instead of a digital pen you, as an artist, have created a digital artwork, you wish it to be attributed to you and to receive compensation for its creation, how can you do so? You could possibly use the traditional licensing of such digital art, which often relies on some sort of physical entity, such as a certificate, limited edition prints, signed licences[13]; but these do not always work well so you may resort to blockchain technology.

Ascribe and Verisart

The first approach to addressing the problem of licensing *digital art* was made by Ascribe, a company established in 2014, which uses blockchain.[14]

Ascribe enables artists to create unique copies of their digital works and offers them a platform for uploading them, securing their sharing, selling and attribution.[14] The artworks can be sold with straight-forward licensing and allows for the easy tracking of successive transfers of ownership. To maximize virtual security, the blockchain is encrypted with a specific key.

Market places, collectors and galleries can consequently rest assured that they have acquired *authenticated* digital works.

As pointed out by Johnson and McNamara: "The work is time stamped on the blockchain. It's authentic because the provenance is clear on the blockchain and always links back to the original creator."[15]

But *digital art* is only a small segment of the art market, and to bridge the gap between *physical art* and blockchain, a new company Verisart was founded in 2015[1,16] by Robert Norton (Fig. 43).

At present, Verisart provides, by invitation, a free app and website to renowned contemporary artists and collectors so they may generate certificates of authenticity. The artist takes a picture of the work, adds its title and dimensions, the materials used and year of production and signs off like a normal certificate. The certificate is then given a URL allowing verification of provenance, as well as a cryptographically

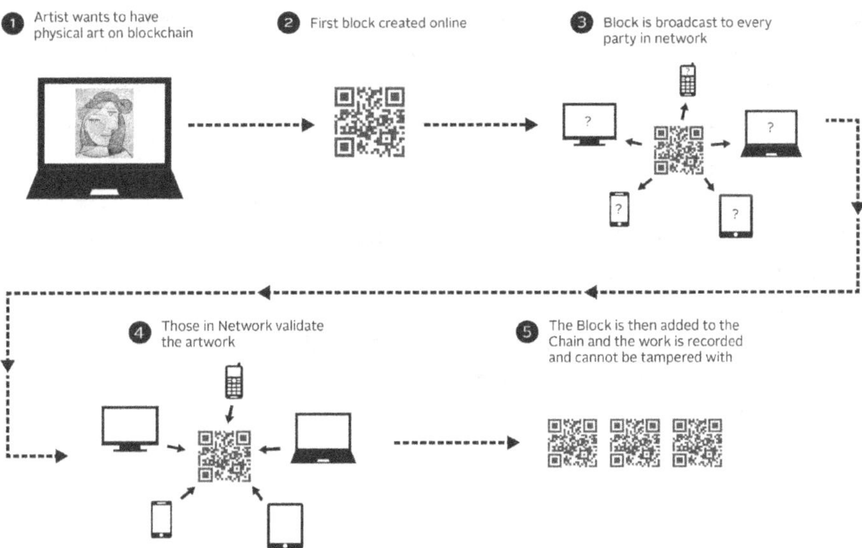

Fig. 43 How blockchain works for physical art

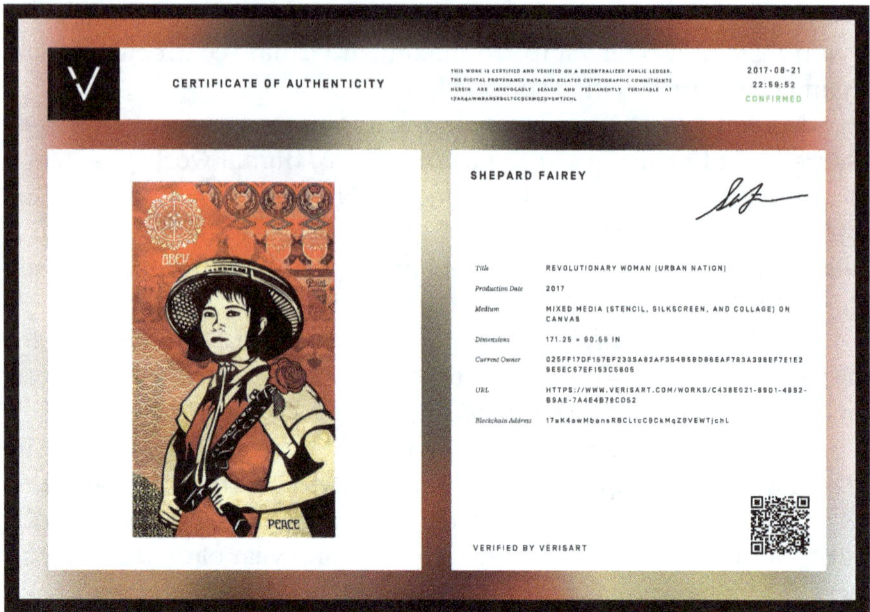

Fig. 44 Certificate of authenticity verified by Verisart

secure registry, which is time-stamped. Such a certificate (Fig. 44) fulfils the international standard for describing antiquities and art (i.e. Object ID compliant) and is a lasting record that can be shared and circulated[1,16] based on the blockchain, which as we know is a decentralised tamper-proof database.

The certificate benefits *collectors*, who virtualize their assets without having to reveal to all price and identity; *artists*, who gain a free inventory management system and *galleries*, who are informed when an artwork changes owner.[16]

Such a system will enable users to verify a work's authenticity, condition, and provenance from their mobile phones before engaging in any online auction or sale.[11]

There is no doubt that in the very near future, *blockchain technology* will revolutionise the art world.

Part III

La Bella Principessa

A Possible Solution to the Enigma

When a thing ceases to be a subject of controversy, it ceases to be a subject of interest.

William Hazlitt[1]

A vellum sheet portrait catalogued at Christie's as being 19th century and possibly German, was sold in 1998 to Kate Ganz, a New York art dealer for $21,850. Nine years later, the portrait was resold to the American dealer Peter Silverman for the same price and with the same ascription.[2]

The portrait is that of a beautiful young lady who appears to belong to the Milanese nobility.[3] This striking profile of a young girl almost speaks to and moves us, not only on account of the beauty and delicacy of its execution but also by way of the reflective expression in her eye (Fig. 45).

The drawing drew the attention of Alessandro Vezzosi (Italian Art Critic and Leonardo Scholar), Nicholas Turner (Former Curator of the British Museum and the Paul Getty Museum) and Cristina Geddo (expert in Milanese Leonardesques), who attributed it to Leonardo, an ascription which would be followed by both Martin Kemp (Emeritus

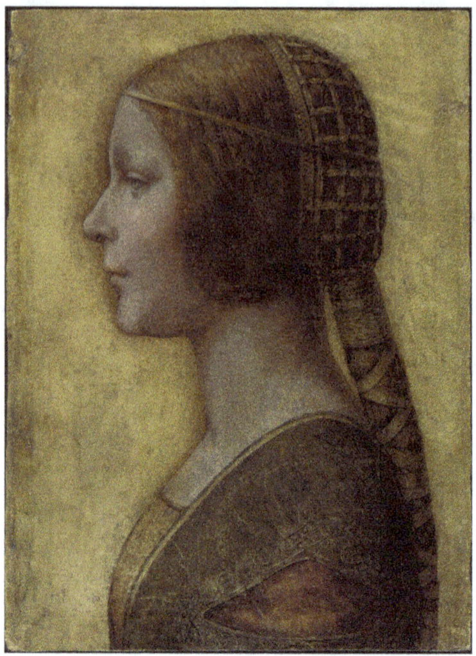

Fig. 45 *La Bella Principessa*

Professor of the History of Art, Oxford University and renowned authority on Leonardo da Vinci) and Pascal Cotte (French Scientist, Chief Executive Officer and President at Lumiere Technology).[3,4]

Favourable endorsements for such an attribution followed from a number of Leonardo and art experts, but scepticism was also expressed by a number of Leonardo scholars and art connoisseurs.

Since the artwork obviously belonged to a vellum book, the biggest challenge for Martin Kemp was to find such a volume — a task which at the outset seemed virtually impossible. At the suggestion of art historian D.R. Wright,[5] Kemp embarked on the search for one of the books entitled the *Sforziada*, which was specifically made for the wedding of Bianca (daughter of the Duke of Milan Ludovico Sforza and his mistress Bernardina de Corradis)[3] and which was housed at the National Library of Poland. Wright suggested that the version at the National Library of Poland[3,5] might be of particular interest.

Kemp and Cotte travelled to Warsaw and using Cotte's macrophotography (giving extreme close-ups), they discovered that on examination of the folios, that a folio had been removed from the Sforziad vellum book.[6]

A close scientific analysis was carried out by Pascal Cotte, the results of which were discussed in a book published in 2010 by Cotte and Kemp entitled *La Bella Principessa: The Story of the New Masterpiece by Leonardo da Vinci.*

As related in their book, three small holes in the left margin of the artwork were found to line up with the original binding of the Sforziad and the dimensions of both the folio and portrait were found to be very close.[3]

Amongst the scientific evidence put forward and pointing towards authenticity was that obtained by multispectral and infrared imaging as well as by C^{14} dating (pages 158, 154, 147).

Multispectral imaging revealed a distinctive hatching in which the strokes are inclined towards the left at an angle close to 45°[3] indicating that the artist was left handed, as was Leonardo.

The infrared images also obtained through multispectral imaging indicated distinctive *pentimenti* which had adjustments reminiscent of the *pentimenti* observed in Leonardo's *silverpoint drawing of a Woman in Profile* at Windsor. As for the C^{14} dating using AMS (page 146) carried out by using a tiny sample of vellum taken close to the edge of the painting, it revealed with "95.4% probability [...] a bracketed date of 1450–1650 AD"[3] indicating that the vellum came from an animal deceased around AD 1450 and thereby confirming that this work could have been made during Leonardo's lifetime.[3]

The initial concerns about such an attribution pointed to the fact that Leonardo never drew on vellum! Such reservations were subsequently rebutted, indicating that the sixty illustrations provided by Leonardo for Luca Pacioli's (ca. 1445–1517) *De Divina Proportione* are all on vellum, in ink with coloured washes (same as in *La Bella Principessa*). Furthermore, Leonardo always liked to experiment with new techniques and was specifically interested in the question of "dry colouring" in coloured chalks.[7,8]

Such counter-arguments would assuage some but not all in the art world!

It is indeed interesting that since the publication of Kemp and Cotte's book in 2010 to the present day, no work of art has stirred as much polemic as that of the attribution of *La Bella Principessa* to

Leonardo. Arguments have been put forward questioning, in addition to vellum as the support, the credibility of one of its earlier owners Jeanne Marchig, the evidence of *pentimenti*, the dimensions of the artwork, the number and position of the stitch holes as well as the style and execution of the work.[9–14] Many of the arguments put forward were refuted by Martin Kemp.[15,16]

Media coverage reached a crescendo when Shaun Greenhalgh a notorious forger, stirred the highbrow art world, by claiming in his book, *A Forger's Tale*, that he was the author of *La Bella Principessa*.[17–25]

Moratorium on Further Discussions

Martin Kemp, consumed by all the controversy surrounding him as a result of his attribution expressed with a perceptible tinge of sadness:

> *I wholly reject the personalisation of the debate. A matter of professional judgement is just that and no more. I would rather be right than wrong, but no one has a divine prerogative to be right.*[26]

He then announced that he was putting a personal moratorium on any further discussions on the case unless new evidence turns up. "I will not be saying more on this unless genuinely new evidence comes to light."[26]

But can new evidence be brought to light? And how true is Greenhalgh's assertion?

Shall We Ever Know Her True Story?

The authenticity of a painting is generally difficult to establish when it displays features that seem uncharacteristic of the artist. Generally, strong debates ensue between experts with opposing views, and authenticity can only be conclusive in the presence of *incontrovertible* evidence.

Such was the case of van Gogh's *Still Life with Meadow Flowers and Roses*, mentioned earlier, which was considered uncharacteristic of van Gogh because of the unusual size of its canvas (100 × 80 cm), the position of the signature and the large number of flowers, quite uncommon features in van Gogh's still lifes. The *irrefutable* confirmation needed for establishing authenticity was brought about as we have seen by one single experiment carried out by synchrotron X-ray imaging (page 152).

In the same vein, the painting of the *Infanta Margarita* mentioned in (page 51) was identified by different connoisseurs in the 1970s as a copy after Velázquez dating 1850–1870. However, some experts attributed the work to Manet. Here, again the *incontrovertible* evidence came by means of experiments by polarised light microscopy (page 137), through a comparison of the configuration and nature, of the pigments in two of Manet's certified paintings *Spanish Ballet* and *Woman with a Jug*, with the pigments in the painting in question.

Similarly, experts had been arguing for decades about the authenticity of *Saint Praxedis* a painting attributed by some to Johannes Vermeer. Reservations had been strongly voiced because of its unusual religious character with regard to Vermeer's secular oeuvre. The *incontestable evidence* and subsequent authentication came from the recent technique of lead isotope ratio (page 146) through the comparison of minute amounts of lead white taken from both *Saint Praxedis* and from another contemporary and uncontested work by Vermeer, *Diana and her Companions*.

The question on *La Bella Principessa*'s authenticity could be addressed by performing a similar lead isotope ratio experiment as that carried out on *Saint Praxedis*. Comparison of the location of the ore from which the lead (Pb) in *La Bella Principessa* was extracted with the ore used as a source of lead in any of Leonardo's paintings or drawings created around the same time, (such as *De Divina Proportione* or the *Lady with an Ermine*) could help provide closure to the drama. If, in addition, the lead white used in LBP is found to be identical with that in any of these Leonardo creations, this would provide the needed *irrefutable evidence* of authenticity.

The crucial question remains however: would any museum or library approve the extraction of minute amounts of lead from one of their treasured Leonardos to allow the solving of one of the important enigmas in the art world, that of *La Bella Principessa*?

Greenhalgh's Dubious Assertion

How many claims have created more ripples in the art world than that made by Shaun Greenhalgh as to his faking of La Bella Principessa?

Shaun Greenhalgh, born in 1960, is the British art forger mentioned earlier (page 13), who, between 1989 and 2006, produced a wide range of forgeries. His elderly parents and brother dealt with the sales side of the sham with his wheelchair-bound father convincingly carrying out most of the transactions while acting as an innocent invalid.

A misspelled cuneiform on a forged 7th century BC Assyrian relief tablet uncovered the whole scam and led to Greenhalgh's arrest in November 2007, and to a prison sentence of four years and eight months.

Close to five years after his release from prison, Greenhalgh claimed in his book *A Forger's Tale*, that he was the author of *La Bella Principessa*.

He wrote:

> *I drew this picture in 1978 when I was at the Co-op* [...] *The 'sitter' was based on a girl called Sally who worked on the checkouts. Despite her humble position, she was a bossy little bugger and very self-important.*[17,18]

Greenhalgh disregarded the many elements that deprive his assertion of all credibility. In his claim, he was totally oblivious to the fact that *La Bella Principessa's* provenance could be traced back to at least 1955 when, in a lawsuit brought in 2012 by Jeanne Marchig against Christie's, she indicated that the drawing belonged to her husband artist Giannino Marchig when they married in 1953.

Moreover, in 1978, Greenhalgh was only 17 years old, it is unlikely that at such a young age he would have become aware that Leonardo's left handedness had to be emulated. And if Greenhalgh did indeed forge LBP, why did he remain silent for so many years and why did he not promote the drawing as a Leonardo?

There are many inconsistencies in Greenhalgh's notification, one scientific observation in particular cannot be disregarded and it is that the white pigment used in this drawing contained lead which a reliable laboratory dated back to at least 250 years[25] through the mass spectrometric determination of lead isotopic ratios (page 146). Greenhalgh's response that he was using organic pigments challenges his credibility even more since lead as lead carbonate is inorganic and could not have been present in any organic medium.

So why did Greenhalgh lie?

According to criminologist Maurice Gauthier, "the act of forgery is motivated [...] by inner conflicts that demand relief from tension [...] an attempt to solve emotional conflicts."[27] In keeping with Maurice Gauthier's conclusions, psychiatrist Charles V. Ford, in his book *Lies, Lies, Lies, The Psychology of Deceit*, identifies compulsive lying as a coping mechanism that helps in alleviating the individual's internal stresses.[28]

The predilection for lying has already been observed in the case of a good number of famous forgers, such as the likes of Han van Meegeren, Elmyr de Hory and Eric Hebborn.[29]

In his compelling story on van Meegeren *The Man who Made Vermeers* Jonathan Lopez[30] reveals an individual with extreme antisocial attitudes, a crook and a sociopath, which was a far cry from the legendary Dutch folk hero that the art world represents. An inclination to lie as well as an interest in forgery can be detected as early as 1913, when at the age of 24 or so, a number of years before his notorious forgeries, van Meegeren made a copy of his *Study of the Interior of the Church of Saint Lawrence* (for which he won a Gold Medal from his school in Delft) and intended to sell the duplicate as the original to a second buyer, but was later dissuaded to do so by his first wife Anna de Voogt.[31]

Elmyr de Hory's predilection for lying is uncovered in the memoir of his former live-in partner Mark Forgy,[32] who suggested that he lied to his first biographer Clifford Irving[33] when he claimed that his father was an aristocratic Austro-Hungarian Ambassador and his mother belonged to a family of bankers. In true fact, records retrieved at the Association of Jewish Communities in Budapest listed his father as a wholesaler of handcrafted goods and his mother as a holocaust survivor.[34]

Similarly, Eric Hebborn's tendency to deceive was revealed by his former partner Graham Smith when he refuted Hebborn's claim of having redrawn a damaged Leonardo cartoon at the National

Gallery, London.[28] Hebborn's statement that he had painted a Rogier van der Weyden and an Annibale Carracci were also exposed as a lie.[28]

Greenhalgh's claim may well have stemmed from a desire to achieve greater notoriety, but it is perhaps even more likely that the underlying cause for his action was a need to relieve inner conflicts through an urge to lie.

As the enigma of *La Bella Principessa* continues to baffle, do we have enough reason to hope that we shall one day know her true story?

Mona Lisas

Imitation is the sincerest form of flattery that mediocrity can pay to greatness.

Oscar Wilde[1]

In the entire history of art, no work has ignited greater curiosity, aroused more emotions than the Louvre's *Mona Lisa*, and none has made more headline news or been surrounded by such an aura of mystery.

Innumerable versions have been made of this beautiful 'Lady Lisa,' some as legitimate copies (Fig. 46), others as individual reinterpretations and others still, purported to be authentic second creations by Leonardo himself (Fig. 50).

The Louvre *Mona Lisa*

Today, it is only the Louvre *Mona Lisa*, this smiling sphinx as described by Théophile Gautier, which is the uncontested authentic version, attracting millions of admiring onlookers (Fig. 47).

As eloquently expressed by art historian Kenneth Clark: "... this strange image strikes at the subconscious with a force that is extremely rare in an individual work of art."[2]

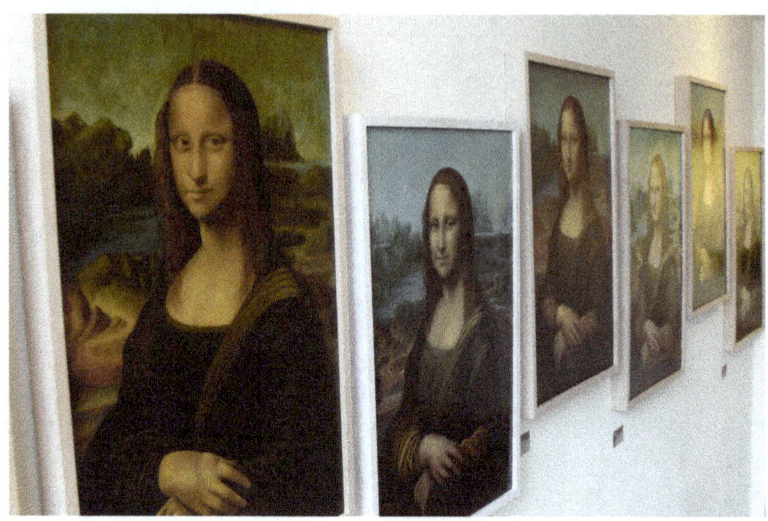

Fig. 46 Copies, copies, copies

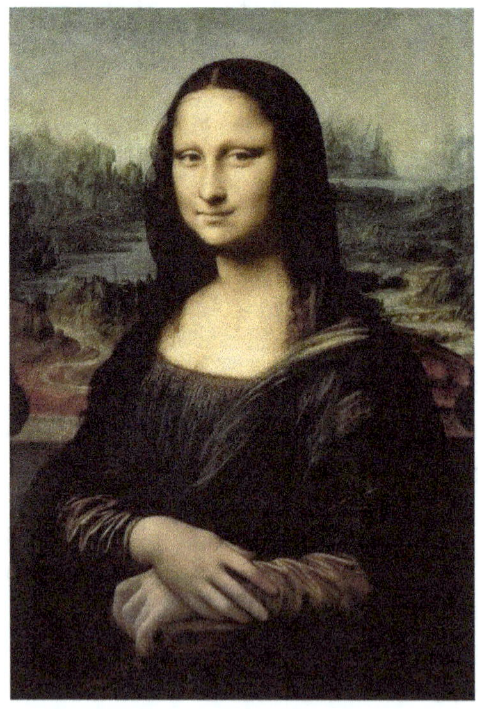

Fig. 47 The Louvre *Mona Lisa*

In this portrait believed to be that of Lisa Gherardini, wife of the rich Florentine silk merchant Francesco del Giocondo,[1] Mona Lisa gazes at us with an enigmatic smile that alternatively appears and disappears before our stare. Leonardo's genius in executing effects of subtle illumination and exquisite softness of lines renders her seemingly happy one moment and mocking the other. The dark veil covering her hair is believed to have been commonly worn to reflect virtuous behaviour, whereas her plain clothing remains one of the riddles of Leonardo's portrait as it is quite different from what one would expect for a woman married to a prominent businessman in Florence.[3,4]

Part of Mona Lisa's notoriety goes back to 1911 when news of her disappearance was received with disbelief and denial. An immense commotion ensued. Jean-Pierre Cuzin the former curator of paintings at the Louvre recalls:

> *The public came just to see the void where the painting had been hung, just to see the nails which held her. Everyone thought that she was lost forever. She was a national treasure! There was a huge uproar. It was a major event.*[5]

And as related by the late writer and cartoonist Seymour Reit:

> *Then, of course, the French temperament took over and they began to have fun with it. There were jokes. There were riddles. There were cartoons. Somebody wrote to the newspapers and said, 'When are they going to take the Eiffel Tower? That's obviously gotta go.' They printed sheet music about the theft of the Mona Lisa, which they sang in cafés. There was a chorus line in one of the cabarets that came out all dressed as the Mona Lisa.*[5]

Suspicion grew to such paranoia that the first suspect to be formally questioned was French writer Guillaume Apollinaire, followed by Pablo Picasso. In the end, no charges were brought against Picasso; Apollinaire was arrested, but as no evidence surfaced, he was released in less than a week.[6]

In 1913, the culprit Vincenzo Peruggia was finally caught when he tried to negotiate an agreement with a Florentine art dealer of good repute and with the director of the Uffizi Gallery in Florence.

When interrogated, in the former Murate (walled fortress) prison in Florence, on December 15, 1913, Peruggia described the theft as follows:[4]

> *I have been working in the Louvre on two occasions, the first in 1909, the second in 1910, and remained there three or four months each time. I helped clean the canvases and put them under glass.*
>
> *On a morning of August […] about 7.30, dressed as a worker with a white linen smock, I went to the Louvre, and I entered, through a back door where workers were entering, a door that opens towards the Seine. I followed in and found myself […] in the Square Salon [the Salon Carré] and my choice fell, with no premeditation, on the Gioconda by Leonardo da Vinci.*
>
> *The picture of the Gioconda was fixed on the wall on two hooks and I did nothing more than raise the picture and detach it. Then I hid in a small stairwell adjoining the Room of the Seven Meters [known from the height of its ceiling] — in a corner of it — and with a screwdriver (which I had brought expressly with me) undid the screws that secured the picture and took it out, leaving the frame in the stairwell, and finally I descended into the courtyard where I was hiding the picture under my smock.*
>
> *The custodian, who was busy at the exit door with others, did not notice me […] I do not know the name of the custodian. I returned immediately to my house (Voie de l'Hospital Saint Louis no. 5) and I left the picture in my room without hiding it….*[4]

According to Kemp, "Peruggia exaggerated the ease with which he exited the Louvre. He found the door at the bottom of the staircase firmly locked. He seems in desperation to have unscrewed the door knob, to no avail. Fortuitously a worker arrived and opened the door. Peruggia looked like a legitimate employee."[4]

Peruggia was convinced that Napoleon's army had stolen the *Mona Lisa* and that his theft would be perceived as an act of patriotism since the masterpiece would return back to Italy where it really belonged.

He would later become a kind of folk hero. When put on trial, he maintained that his motivation was of a purely nationalistic nature[5] and got off with a relatively light sentence.

The retrieval of the *Mona Lisa* allowed it to grace the Louvre once more.

Although the theft did much for the painting's fame, no one ever denied Leonardo's exceptional artistic mastery and genius in creating it.

Glaze Technique

Leonardo's *sfumato* (meaning smoky in Italian[7,8] — a technique where contours merge imperceptibly with their surroundings as if they were smoke) has long attracted a great deal of attention. Experts were intrigued by how he had managed to produce this smoky effect which led to the blurring of the contours of her face and an aura of mystery to her eyes.

Scientific studies leading to a better understanding of *sfumato* were previously scarce as there was no non-destructive in situ approach that could allow for the individual determination of the composition of each paint layer.

It is only during the last decade that cutting-edge scientific techniques were applied to address some of these queries.

Two successful approaches recently achieved such a goal. One entailed Philippe Walter et al's experiments using XRF spectrometry (page 149), in conjunction with a specially designed new software, which allowed them to determine the nature, composition and thickness of each layer. [7]

The other was Pascal Cotte's multispectral imaging working in tandem with Professor Mady Elias's pioneering method of pigment identification (page 157), which provided useful information on the pigments used in the different layers of the painting.[8]

Philippe Walter and Co-workers

In 2010, French scientists from the Center of Research and Restoration of the museums of France in a collaborative[7,9] project with the European Synchrotron Radiation facility in Grenoble arrived at what seemed to be the impossible. Using a high technology X-ray fluorescence spectrometer in conjunction with a specialised software, they analysed seven different faces in paintings by Leonardo. In the artworks where glaze (a binder in most cases made of linseed oil tinted by a certain amount of pigment; in Leonardo's case, the amount was generally minute) was used, they succeeded in determining one by one the composition and thickness of the many layers (sometimes up to 30, half a hair thick).

Results obtained on the different analysed faces showed Leonardo's propensity for trying out new techniques, using glaze in some cases and not in others, manganese dioxide in some shading and copper in others, or simply adding a covering of dark pigments as customary in oil paintings.[7]

In the case of the *Mona Lisa*, this investigation showed that a glazing technique was used and that dark areas were obtained by a thicker application of a manganese and iron-containing layer, with underlying layers containing lead white of equal thickness throughout. His addition of films made of an organic medium permitted the formation of approximately 30 translucent layers, each a few micrometres thick. Each layer had to be dried before proceeding onto the next, the multiple layers contributing to the natural look so characteristic of his paintings.

Reaching so many translucent layers is exceedingly complex and time-consuming. Deeply impressed, Philippe Walter, one of the principal investigators in the research, dubbed it: "an amazing achievement even by today's standards."[7]

Cotte and Elias

Using the technique of *multispectral imaging* (page 158) (Fig. 48) and focusing on *Mona Lisa*'s face, Cotte and Elias also tried to clarify Leonardo's s*fumato* technique. They identified a unique umber pigment in the various stratified layers. Iron oxide with a minimal amount of manganese was found to form the upper layer and subsequent layers, whereas the base consisted of 1% vermillion and 99% lead white.[8]

Leonardo, they concluded, used a glaze technique that would give a naturalistic flesh-like appearance reminiscent of techniques employed by Flemish Renaissance painters.

Although these results concur to a certain degree with those of Walter et al., the latter suggested a slightly different approach to that

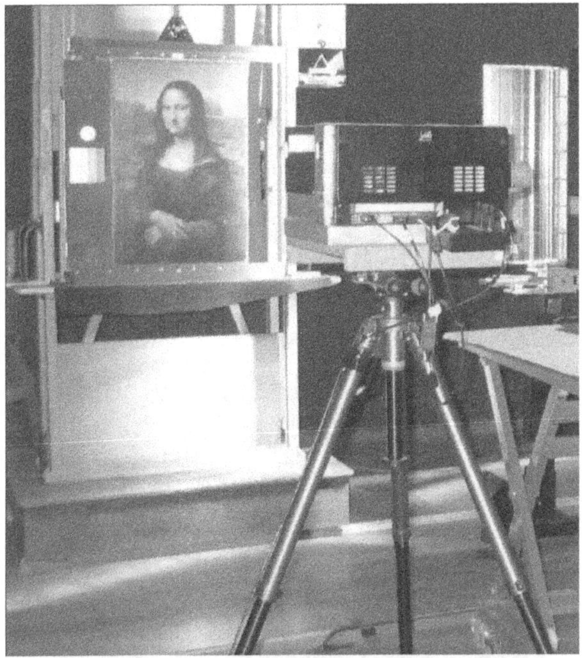

Fig. 48 Multispectral imaging of Leonardo's *Mona Lisa*

of the Flemish Renaissance painters in the way Leonardo darkened the flesh tones.[7] Both the techniques used by Walter et al., and by Cotte and Elias played key and complementary roles in pigment identification and in unravelling Leonardo's glazing technique.

Cotte's multispectral imaging on the *Mona Lisa* would reveal further astonishing information.

Astounding Discoveries on the Louvre Mona Lisa

In addition to pigment identification, for more than a decade, Cotte analysed his 2004 multispectral imaging of the Louvre *Mona Lisa*. His study revealed in the underlayers another portrait of a woman totally different from the *Mona Lisa* we know. Quite unlike the famous 'Lady Lisa', there was no smile on her face, and her sideways gaze was somewhat blank (Fig. 49).[10–14]

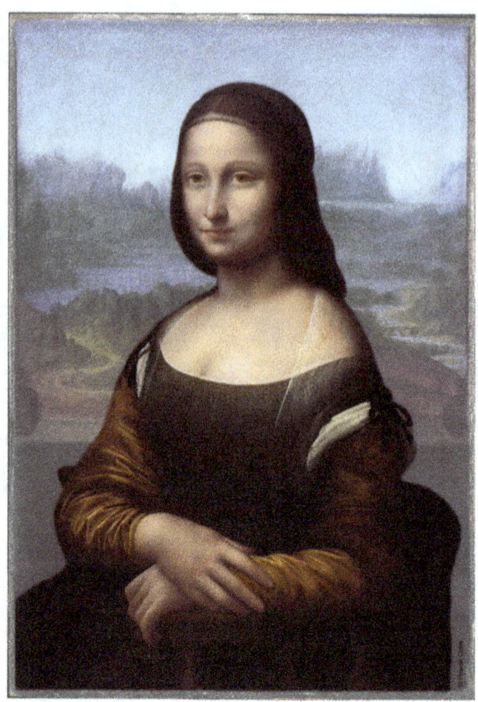

Fig. 49 Hidden portrait found by Cotte under *Mona Lisa*

An additional blurred outline of another portrait as well as a Madonna-type rendition were also observed by Cotte. According to him, such a discovery puts into question Mona Lisa's identity and alters "our vision of Leonardo's masterpiece forever."[10,11]

Not surprisingly, the revelation was met with exclamation in the art world. Leading art historian Graham–Dixon dubbed the discovery "definitely one of the stories of the century" and exclaimed, "what we're talking about is goodbye Mona Lisa, she is somebody else."[11,12]

Art critic Jonathan Jones decried,

> *"Why did the painting change so much as it was worked on? Did Leonardo in fact reuse his picture and paint someone new on top? No. There's no reason to think that."*[13]

Martin Kemp, Emeritus Professor of the History of Art at the University of Oxford, while recognising the ingenuity behind Cotte's images in casting light on Leonardo's train of thought, is "absolutely convinced that the Mona Lisa is Lisa [...] the idea that there is that picture as it were hiding underneath the surface is untenable."[11,12]

In a more elaborate explanation of his point of view, Kemp wrote:

> *Pascal's mode of analysis, adapting mathematical techniques from signal processing, is revealing far more from the deeper layers than was previously possible, but it does not definitively isolate information from a single layer. We are also unclear as to what is happening as the different frequencies of light penetrate the paint layers to varying degrees and interact in diverse ways with the varied optical properties of the materials within the layers of the picture. There is always the danger of seeing what we want to see. None of us are immune from this. Pascal is opening up very important fields for analysis. We are at the beginning [...] Anyone is unwise to pronounce with certainty at this stage.*[15]

For all the expressions of consternation elsewhere, the news was received by the Louvre Museum and by numerous art historians with a wall of silence. More than a decade ago, the Museum had granted Pascal Cotte access to the painting, but had not participated in any of these investigations and declined to comment all along.

Will this famous "Gioconda" continue to defy our expectations and keep us marvelling, as it has long done?

The Isleworth *Mona Lisa*

A slightly different Mona Lisa gazes at us with her cryptic half-smile. She looks very much like a younger and fresher version of her Parisian rival, yet one can perceive subtle differences in the features suggesting that it may be a copy or possibly an altogether different creation (Fig. 50). Had she too been painted by Leonardo or did a gifted artist try hard to emulate the master's style and add discreet variations?

The subject of our intrigue is the Isleworth *Mona Lisa* purchased in 1913 by Hugh Blaker an English painter, art collector and connoisseur from Isleworth in West London.[16] At Blaker's death in 1936, the painting was passed on to his sister Jane and then after her death in 1947, it was bought in 1962 by the American art dealer and publisher Henry Pulitzer in partnership with a Swiss syndicate.[17]

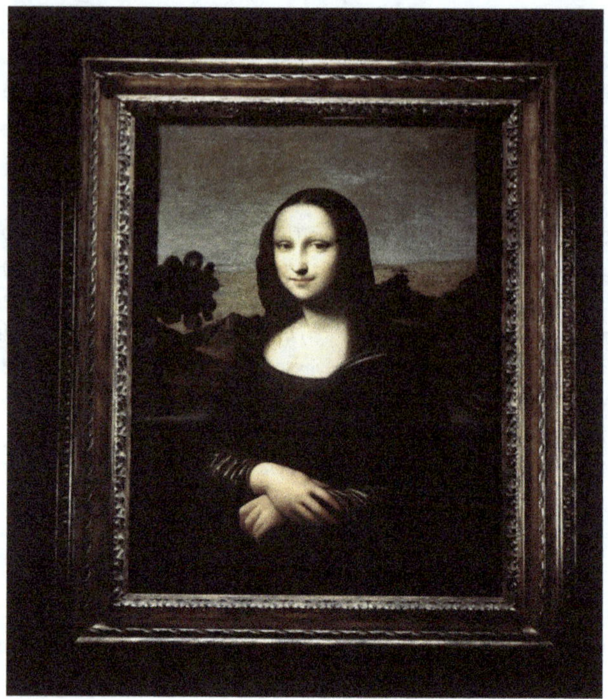

Fig. 50 The Isleworth *Mona Lisa*

Pulitzer was so convinced that it was an original *Mona Lisa* that he sold his Knightsbridge home with its contents to buy the painting in the 60s, but subsequently never managed to convince the experts of its authenticity.

Philatelist and auctioneer David Feldman became involved when, after Pulitzer's death, the painting passed to his Swiss wife and after her death to an unknown Swiss consortium of individuals who deposited the artwork for 40 years in the vault of a Swiss Bank. The consortium established the Mona Lisa Foundation (MLF) in 2008 with Feldman as vice-president, its main purpose being to research and investigate the Isleworth *Mona Lisa*.[16]

Multispectral Imaging

The MLF sought Pascal Cotte's expertise to undertake the task of analysing the Isleworth *Mona Lisa* by multispectral imaging.

The Foundation reported that Cotte's results indicated that all the pigments were available in Florence during the early 16th century with no detected anachronisms. It was also revealed that the technique used was not a glazing one as in the case of the Louvre *Mona Lisa*. The MLF report on the painting also indicated that Cotte had remarked several times *"the hands are extraordinarily beautiful,"* a statement which was later denied by Cotte.[18]

Adding fuel to the controversy, referring to a two hour press conference organized in September 2012 by the MLF to announce the attribution of the painting to Leonardo, Nick Miller reported:

> *Cotte was not at the September press conference, and The Huffington Post online news site reported that he neither thought, nor asserted, that the painting was a Leonardo masterwork.*
>
> *The Huffington Post also reported that [...] Cotte had been quoted without permission.*[16]

Two Mona Lisas?

In their book *The Mona Lisa Myth* published in 2013, art historians Jean-Pierre Isbouts and Christopher Heath Brown suggest that there is historical evidence of the painting's authenticity: back in 1550, Leonardo's biographer the historian Giorgio Vasari had referred to the *unfinished* Mona Lisa in his book *Lives of the Most Excellent Painters, Sculptors, and Architects*. In addition, he had described some details in her eyebrows which cannot be perceived in the Louvre *Mona Lisa*.[19] With regard to the question of eyebrows, however, multispectral imaging has confirmed their presence in the original Louvre *Mona Lisa*.[17,19]

Historians find further proof in mention by 16th century art theoretician Gian Paolo Lomazzo that Leonardo painted the portrait of *La Gioconda* and of *Mona Lisa*. Many counter-arguments have been put forth, however, refuting the existence of two Mona Lisas.

Martin Kemp has been behind some of these: "It's a perfectly honest, well-made early copy," he wrote of the Isleworth *Mona Lisa*. "Pictures were copied [because] you couldn't go to the Internet and order a reproduction. So if you wanted something like that [the *Mona Lisa*] and you couldn't get a hold of a Leonardo, you would order a copy."[20]

Kemp also pointed out a large list of stylistic details which are inconsistent with Leonardo's style and that: "Infrared reflectography and X-ray do not reveal any characteristics of Leonardo's preparatory methods."[20]

Art historian Donald Sassoon claimed that there is no real evidence for the 'two Lisas hypothesis' but that the idea has its own charm.

Expressing concern, art historian Jan Blanc referred to the wide spectrum of scientific investigations carried out on the Isleworth *Mona Lisa* sponsored by the Mona Lisa Foundation and pointed to the undue importance given to results by scientific analysis and to the error of deeming them to provide irrefutable proof.[20]

One important concern by a large number of art historians and connoisseurs relating to the Isleworth *Mona Lisa* is that Leonardo never painted on canvas but generally painted on wood.

Responding to such claims, the MLF press office indicated that Leonardo had carried out a number of works on canvas while working under Verrocchio in the 1470s.

Dating of the Canvas by Accelerator Mass Spectroscopy (AMS)

The carbon-14 dating of the canvas, commissioned by the MLF, was carried out at the Swiss Federal Institute of Technology in Zurich (ETHZ) and indicated a date between 1410 and 1455 A.D. According to Feldman, this refuted the claims that the painting was a 16th-century copy.

However researcher Irena Hajgas, who carried out the test by AMS at the ETHZ, indicated that the dating was carried out on an unidentified piece of canvas sent to her by the foundation. She claimed that she could only attest to the validity of the dating measurements but could not endorse any of the foundation's conclusions, having not seen the artwork as customary for certification. Furthermore, she claimed that she was informed by the foundation that the canvas belonged to the Isleworth *Mona Lisa* only a few weeks after the announcement of the results, which was also not customary practice.[21]

Her remarks were denied by the MLF press office, which stated that for the carbon-14 dating, the canvas sample was given to ETHZ by a Swiss deputy judge who had clarified where it came from.[21]

Geometric Analysis and The **Vitruvian Man**

Part of the evidence put forward by the Mona Lisa Foundation were the results of Italian geometrist Alfonso Rubino who demonstrated that the geometric principles that Leonardo adopted in his *Vitruvian Man* are also found in both the Louvre *Mona Lisa* and in the Isleworth painting. The *Vitruvian Man* is a drawing from 1490 by Leonardo that represents a man of idealised human proportions, depicted in two overlapping positions with arms and legs outstretched, inscribed in a square and a circle (Fig. 51).

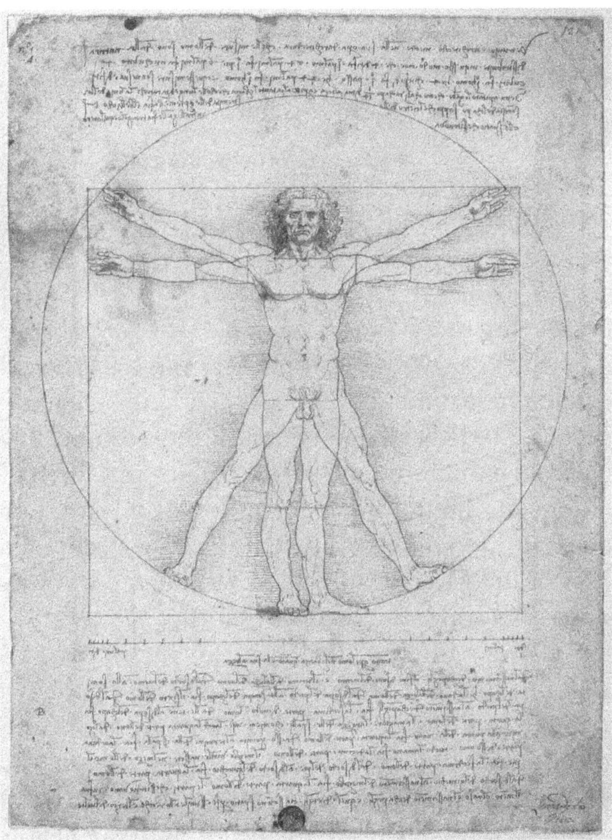

Fig. 51 Leonardo's *Vitruvian Man*

The detection of these analogies did not impress some art historians who argued that such apparent geometry was no longer sought in old paintings and that none of the Renaissance paintings in their various stages of preparation revealed similar constructional geometry.[22]

Alessandro Vezzosi, director of the Museo Ideale Leonardo da Vinci, acknowledged the sense of mystery still surrounding the painting, "I don't want to speak today about the attribution that has been formulated by the Mona Lisa Foundation [...] it's a fascinating possibility [but] a lot of research and technical examinations remain to be done."[18]

The Prado *Mona Lisa*

An Unusual Copy

Ana Gonzalez from the Prado Museum gazed at the painting in front of her and exclaimed,

> *While the corrections are identical, the lines are not [...] Like I write an A and you write an A, you can tell it is not the same [...] in general, it is not nearly so fine a painting.*[23]

She was referring to the Prado version of the Mona Lisa, a painting which would attract considerable attention from the art world (Fig. 52).

Its unusual characteristics became evident when in 2010, the Louvre had requested that it be restored and technically studied in preparation for an exhibition, due to take place in March 2012.

Originally belonging to the Spanish Royal collection, the painting had long been considered to be one of the many anonymous early 16th century copies of the Mona Lisa, and it had been part of the Museum collection since it's foundation in 1819.

Analyses using different techniques including raking light (page 136) and infrared reflectography (page 154) on this artwork covered with varnish and black paint gave unexpected results. Notably, the infrared reflectograms of both this work and Leonardo's original revealed very similar execution steps and a succession of common corrections or *pentimenti*.[24] This led to the initial conclusion that both paintings were created in tandem, presumably by assistants or students of Leonardo, who in keeping with tradition, copied their master's works as part of their training.

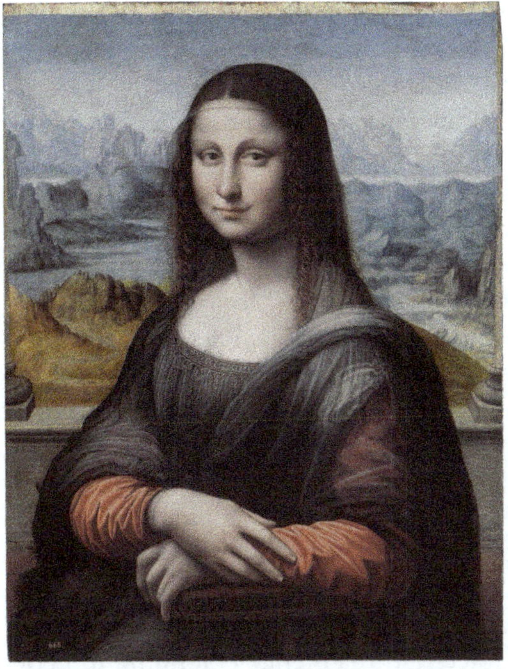

Fig. 52 The Prado *Mona Lisa*

A closer scrutiny of both paintings however showed that the copy was painted on walnut whereas the original was on poplar wood. This important difference led Bruno Mottin[25] to suggest that Leonardo painted the *Mona Lisa* in 1503, probably on the only available wood at the time, which must have been the poplar, and it was therefore quite unlikely that he would work on this coarse wood with his copyist working in tandem on the finer walnut wood.

Gonzalez[24] however attributes the high quality of the materials used in the Madrid painting (including walnut wood) to an important commission.

In spite of the very great similarities between both underdrawings, several details suggested that Leonardo and his copyist probably started to work at the same time with the copyist completing the work at a later stage;[4,25] for instance in the original, the underdrawing showed a different position of the left hand clutching the armrest whereas the copy showed in the underdrawing the left hand in the same position as that observed in the finished original *Mona Lisa*.[25] In addition, the columns and their bases (also a later addition in the finished Louvre *Mona Lisa*) are clearly discernible in the Prado underdrawing.[4]

A landscape with mountains very similar to that in the original painting was also observed beneath the black background. When conservators removed the black paint, a rich landscape was revealed giving clues as to what the landscape in the Louvre *Mona Lisa* (now covered by layers of dust and obscured by darkened varnish) must have looked like.[26]

Now if these two versions of Mona Lisa were to be authenticated, through scientific analysis alone, it would probably be difficult to identify the original through the sole consideration of the data collected on the pigments, the binding materials, the underdrawings *and* the wooden supports. One would expect the scientific investigation to reveal the same palette, a comparable craquelure, identical binding media, closely similar underpaintings and inconclusive results with regard to the supports. In cases like this one, the dry objective data would not on its own be sufficient.

As Byrne writes, only a competent expert eye could have captured some of the characteristic elements, belonging to Leonardo:

> *Like a criminal leaving clues, the painter wants us to identify him, a kind of Freudian slip, manifest in the individualized physiognomic details like ears, toes, fingers, particularly, conspicuous in the 'spontaneous' part of the artistic process.*[27]

In this particular case, where science was unable to show decisively that the Prado *Mona Lisa* is not by Leonardo, only with connoisseurship and a well-trained eye can there be decisive discernment between the original and the copy. From the very outset, the expert would probably perceive in the original painting a decisiveness of execution which would expectedly be lacking in the copy. An evaluation of the technical method that pervades the artwork would then ensue and entail a close examination to name a few, of the brushwork, the colouring scheme and the interplay of the light and colour. More important however is the connoisseur's ability to behold in the original painting the creative soul and mental set of the true master.

Reflections

The Louvre, Prado and Isleworth *Mona Lisas* have been scrutinised by the most up-to-date scientific techniques.

With regard to the Louvre *Mona Lisa*, new light has been shed on Leonardo's *sfumato* technique, and in depth multispectral analysis has revealed the astonishing and controversial existence of an underneath hidden portrait. Such a revelation has been accepted by a precious few, opposed by others with a third large majority remaining silent.

While recognising the tremendous scientific strides and the great potential MSA holds for the future, its limitations cannot be overlooked. The data generated by such a technique is left open to interpretation.

Scientific examination of the two paintings the Isleworth *Mona Lisa* and the Prado *Mona Lisa* did not indicate any anachronistic elements or reveal questionable dating of the supports. However, in the former case, on the basis of the scientific evidence alone, no conclusive attribution could be made supporting the hypothesis that this work was created by Leonardo and reservations have been expressed by a number of art–historians and connoisseurs.

In the case of the Prado *Mona Lisa*, a close stylistic analysis led to the unequivocal conclusion that it was a copy created in tandem with Leonardo's *Mona Lisa*, but completed some time after the original artwork, with connoisseurship playing a crucial role in this attribution process.

Such examples point to both the importance and limitations of science. It has become more evident that for a definitive verdict on authentication there needs to be a consensus between scientists, art–historians and or connoisseurs.

Such a consensus of opinion followed by authentication has been observed in the recent case of Rembrandt's self-portrait, where the evaluation by the distinguished Rembrandt expert van de Wetering was in harmony with the scientific results obtained at the University of Cambridge Hamilton Kerr Institute's laboratory (page 219).

Equal caution must be exercised with regard to categorical evaluations made by the world of connoissership, as we cannot fail to remember the case of van Gogh's *Still Life with Meadow Flowers and Roses* (page 57) where, in spite of the fact that no anachronistic element was detected by scientific analysis, experts had cast doubt on its authorship. A letter addressed by van Gogh to his brother, Theo, later confirmed the authenticity of this artwork.

Alas, in the case of most masters, we do not have such unambiguous art-historical evidence. Scepticism put forward by connoisseurship may also be overruled if the provenance of an artwork, supported by scientific evidence, can be clearly determined.

Works of art are sometimes ascribed to a particular master, then unattributed and possibly once more reattributed with the development of new research approaches. It is noteworthy, however, that in the world of authentication one perceives the general reluctance by experts to pronounce ahead of others whether a particular work is genuine or not. They tend to await the emergence of a consensus opinion and to remain silent until they receive categorical evidence. Scholars are taught to be doubters, with consensus being difficult to achieve if at all possible.

When Can Science Play a Definitive Role in Authentication?

The road so far travelled in the art world indicates that a conclusive verdict *authentic* or *forgery* can only come from science in the following situations: (a) when it acts as a tool for *falsification* — in itself invaluable — through the revelation of anachronistic elements in the painting, (b) when it assists in *authentication,* with a parallel consensus and unified opinion emerging in the world of connoisseurship and (c) when it confirms *authentication* as backed by either art historical evidence or by a documented proof of provenance.

Part IV

The Scientist

Introduction

Scientific Tools of the Art Detective

In what follows, the reader will be introduced to a wide spectrum of old and new scientific methods for the study of paintings, progressing from the simple to the complex and entailing the examination of the surface, frame and body of the artworks as well as looking through paint layers.

The initial approach of the scientist and art historian alike in the evaluation of a painting is the examination of the surface of the artwork, determining if the craquelure (a pattern of fine cracks that appear on the surface of a painting) is genuine, and if the brushwork is in keeping with the artist's style. Ideally, this is carried out through the use of a stereo-microscope. This first step is generally assisted by the use of raking light (light placed at an oblique angle to a painting) which is particularly valuable in revealing the painting's surface texture. Ultraviolet fluorescence provided by a simple UV lamp also reveals the fluorescence properties of the surface, enabling differentiation between old and new additions to the painting.

It is, however, in examining the ground layer and the paint layers that, in recent years, a wide spectrum of improvements have been made to old techniques and where novel and sophisticated methods of analysis have been introduced.

As far as the ground layer is concerned, a standard approach has been the detection of underdrawings by X-ray radiography and infrared

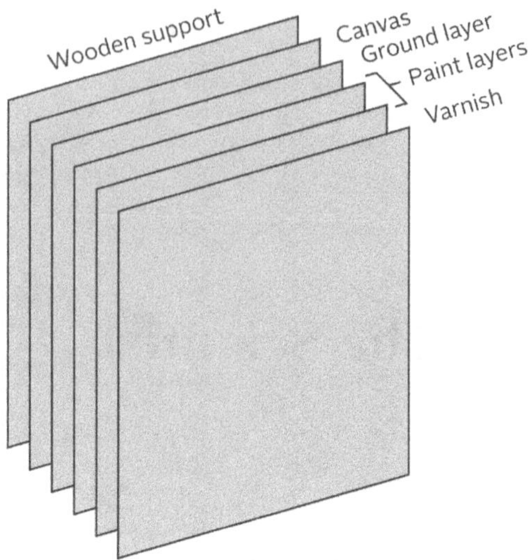

Fig. 53 Different layers in a painting

reflectography. New approaches developed in recent years entail the use of specially designed ultra-sensitive digital charge coupled device (CCD) cameras that in both modes of analysis have resulted in much better resolutions. The use of synchrotron X-ray radiation has also significantly improved images in X-ray radiography. Forgeries are detected through the discovery of anachronisms with the artist's own stylistic development or through the uncovering of an underdrawing belonging to a painter who lived after the purported master's death.

With regard to the body of the painting, in addition to the wide spectrum of available investigative techniques, novel approaches complementing the results obtained by traditional scientific approaches have mushroomed in recent years. Some of these are described in more detail in the following text and include, to name a few, pyrolysis-gas chromatography-mass spectrometry (Py-GC-MS) which has proven invaluable in the analysis of synthetic polymers incorporated as binding agents in paints and proton-induced X-ray emission (PIXE),[1] enabling the *non-destructive* identification of the pigment composition. We may also add the technique of laser-ablation-inductively coupled plasma-mass spectrometry (LA-ICP-MS)[2] which allows the identification of organic pigments and accelerated mass spectrometry (AMS) for the detection of naturally occurring long-lived radioisotopes.

Microscopy-Related Techniques

The expert eye and magnifying glass are instrumental in helping the arthistorian and curator in the initial observation of the surface of a painting. Through the use of raking light, important details on brush-strokes, craquelure, impasto and raised paint are revealed.

However, the information provided by the expert eye, or rudimentary eye-glass is generally insufficient; there is need for additional, more effective methods of examination. A wide spectrum of microscopy-related techniques (some old, others more recent) have been developed and reveal other important complementary information.

Optical Microscopy (OM)

The simplest technique makes use of the OM which magnifies images of small samples and allows the observation of the size of the pigments as well as the cracks on the surface of a painting. The OM is also used to observe the sequence and structure of paint layers.

When applied to the study of paintings, the image is captured on photographic film by a light-sensitive camera. In more modern instruments, a CCD is used to directly display the picture on a computer screen, as in Fig. 54 below.

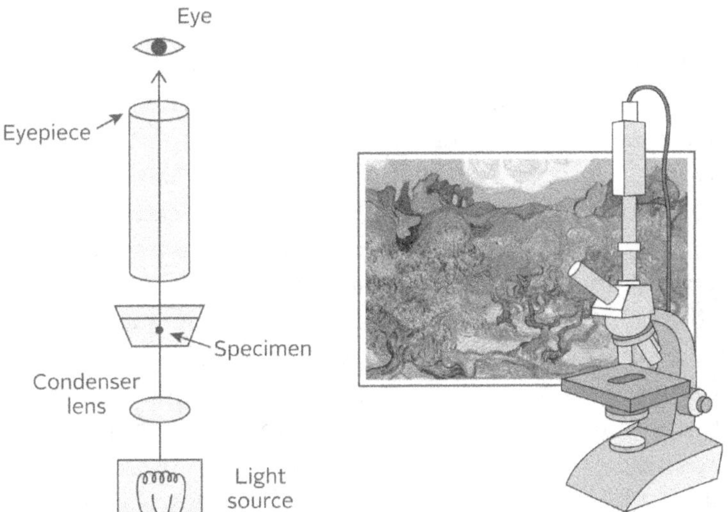

Fig. 54 Principle of an ordinary microscope and microscope with a CCD camera

Stereo-Microscope and Raking Light

A stereo-microscope is an ideal instrument for the examination of the surface of a painting by raking light. This instrument makes use of light *reflected* from as opposed to *transmitted* from the surface of the painting.[3]

Observation by Optical Microscopy of the Craquelure at the Surface of a Painting

Craquelure is a pattern of fine cracks that appear on the surface of a painting and is a sign of ageing of the paint layers due to their shrinking over time. As confirmed by early research work carried out by Spike Bucklow at the Hamilton Kerr institute and by A.J. Varley in the Cambridge University Department of Engineering, "craquelure is a valuable clue in the attribution of paintings."[4,5]

Different styles or prototypes of craquelures can be classified as Italian, Dutch, Flemish and French. Although these cracks can be simulated in a forgery, they can be instrumental in authenticating a painting.

Cross-Section Analysis[6,7]

As paintings generally consist of a canvas affixed to a wooden support which is then covered with a ground layer followed by different layers of paint and an external protective varnish, an OM can be used to observe the layer structure of the paint, giving important insights into the artwork.

Careful study of the cross-sections is usually carried out using an ordinary OM, a UV fluorescence microscope or a polarised light microscope. Information can be obtained on the sequence and structure of the paint layers as well as on the dimension and shape of the pigment grains and their grain size distribution.[6,7]

A fluorescence microscope is essentially an OM with added features. It uses a much higher light source to excite the fluorescent species across the UV–visible light spectrum, then sorts out the weaker emitted light by means of a special filter. This allows the viewer to see only the fluorescent image against a darker background.

When a cross-section of paint is exposed to UV radiation, certain aged organic materials, such as varnish and binders, will fluoresce, emitting light in the visible range at a characteristic wavelength. UV fluorescence microscopy is very useful in the study of materials identifiable as distinctive layers revealing any retouching undergone by the painting, the mixing of colours to produce a particular hue and other details revealing the artist's unique hand.

As to the painter's signature, if there is doubt, a close observation of the cross-section can reveal a layer of dust between the signature and the paint beneath it, which would confirm a forged addition.

Polarised Light Microscopy

A polarised light microscope (PLM) (Fig. 55) is particularly useful for the observation of optically anisotropic pigments that are a few micrometres in size.

Normal sunlight transmits light waves in planes that are all perpendicular to the direction of propagation; however, when the waves are restricted to one plane, the light is said to be plane polarised (Fig. 56).

The polarised light microscope is equipped with a polariser. When placed in the light path before a sample, it allows the sample to

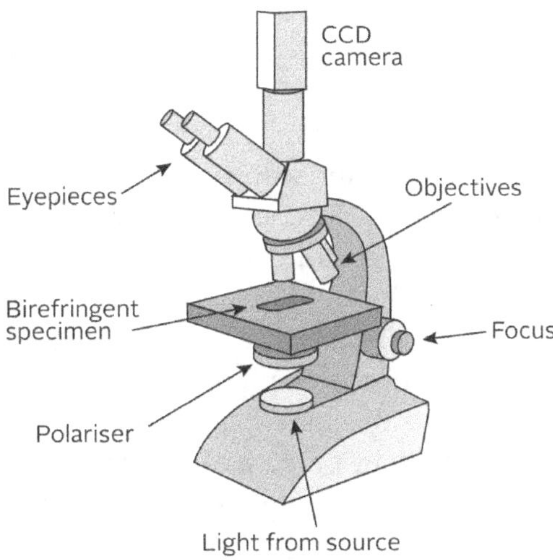

Fig. 55 The polarised light microscope

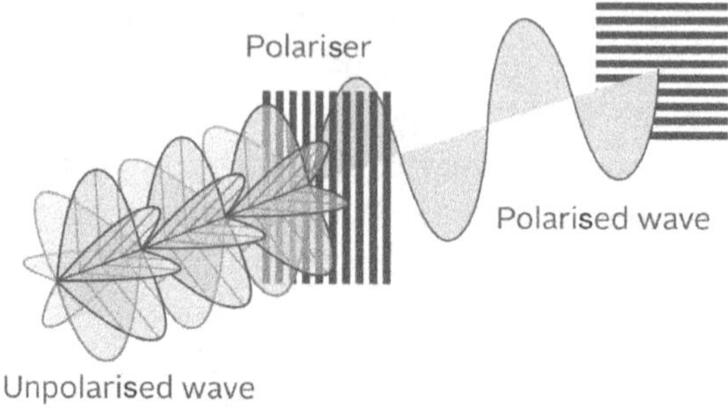

Fig. 56 Polariser and polarised wave

be illuminated with polarised light, as in the figure. If the sample is birefringent (a birefringent sample has two refractive indexes that are perpendicular to each other), it will produce two polarised and mutually perpendicular wave components that are then recombined by an analyser (a second polariser) placed in the pathway above the sample. The pigments can be seen to glow brightly with characteristic colours against a dark background, allowing a clear identification of their colour, size and shape.[8]

Scanning Electron Microscopy (*SEM*)

The composition of each layer in a cross-section of a painting can be determined using a scanning electron microscope. In such a microscope, high-energy electrons, which are thermally emitted from a tungsten filament cathode and accelerated towards an anode, scan the surface of the sample in a raster scan pattern (Fig. 57). In such a process, the electrons interact in different ways with the atoms that constitute the sample and provide information on the sample's elemental composition and topography.

Some of the incoming electrons cause X-rays to be generated, and these act as 'fingerprints' of the elements from which they originated (ED-XRF technique to be discussed later). Other electrons that are nearly normal to the incident path, will bounce back when colliding with atoms in the sample and are generally referred to as back-scattered electrons (Fig. 58).

Regions with different average atomic numbers are detected using back-scattered electrons, as they show a contrast between various

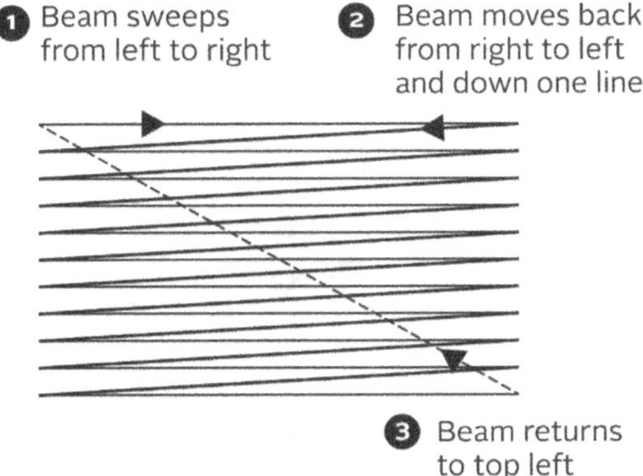

Fig. 57 A raster scan

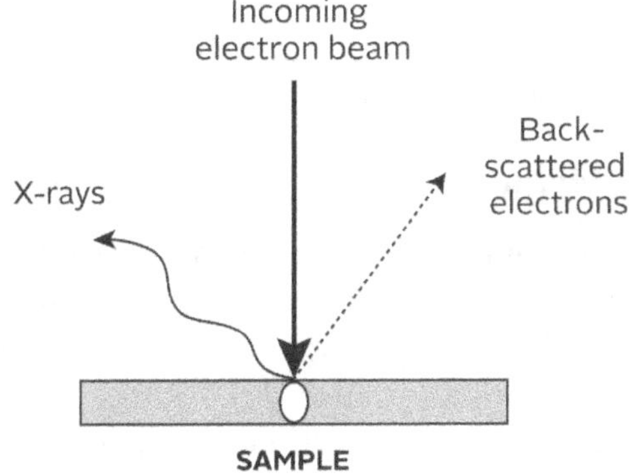

Fig. 58 Generation of X-rays and back-scattered electrons

areas of different chemical composition. In the images produced at very high magnification (up to 100,000 times), higher atomic number elements appear brighter than the lower atomic number ones.

Raman Spectroscopy and Microscopy

Raman spectroscopy has gained attention in recent years as an important analytical tool for the investigation of artworks. The non-destructive nature of this technique makes it ideal for the in situ analysis of pigments, binding media and varnishes. The preferred approach for such an analysis is micro-Raman spectroscopy also referred to as Raman microscopy.

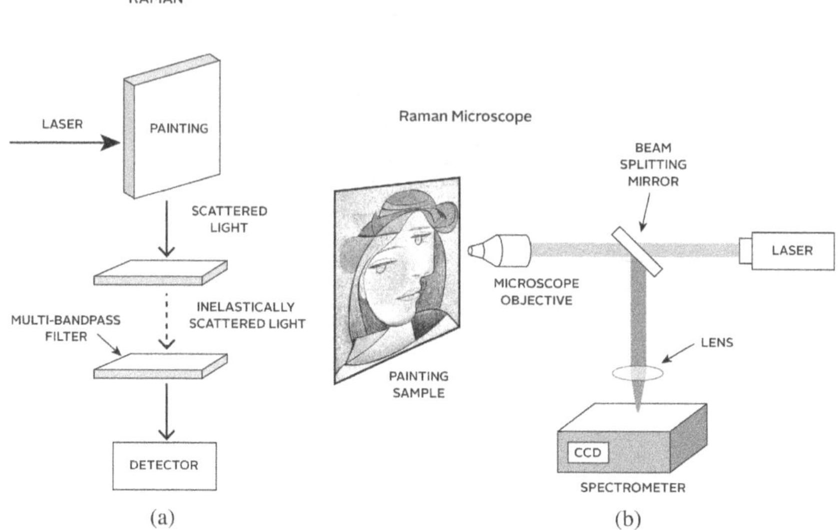

Fig. 59 (a) A conventional Raman spectrometer and (b) Raman microscope

Principle of the Method

Raman spectroscopy entails the interaction of a laser beam (generally in the visible region of the spectrum) with matter. As a result, photons may either be absorbed, reflected or scattered. In the last case, most of the light is generally *elastically* scattered (i.e. it has the same wavelength as the incident light) and a small portion is *inelastically* scattered (i.e. has a wavelength different to the incident light). Inelastically scattered light is referred to as the Raman effect, or Raman scattering, and is very useful in the non-destructive identification of organic and inorganic pigments, as well as binding media (see Fig. 59(a)).

Identification of pigments in Raman spectroscopy is achieved through the comparison of the spectrum of an unknown pigment with that of a reference compound. Inorganic pigments are easily identified, as the spectra of classical inorganic reference compounds are readily available in the literature. The spectra of a good number of synthetic organic pigments have also been determined during the last five years.[9,10]

Raman Microscopy

In their article "Raman microscopy in archaeological science",[11] G.D. Smith and R.J.H. Clark demonstrate the emergence of micro-Raman spectroscopy as an important technique in the analysis of artworks where particles (or pigments) of **less** than 1 micrometre in diameter can be identified. This importance is highlighted elsewhere throughout the literature.[11–16]

The small size of the Raman microscope, its ease of transportation, high sensitivity and its applicability *in situ*, give it great advantages over more conventional techniques. Also, the fact that it is a non-destructive technique makes it very appropriate for museums, particularly in the analysis of valuable paintings and large objects that cannot easily be sent to laboratories (Fig. 59(b)).

Infrared Spectroscopy and Micro-Fourier Transform Infrared Spectroscopy

Infrared (IR) spectroscopy is a widely used technique in a large number of scientific departments belonging to museums. However the preferred approach to the analysis of artworks is micro-Fourier transform infrared spectroscopy (micro-FTIR). This technique is a non-destructive tool for the rapid, effective and easy investigation of very small amounts of samples by allowing the identification of organic pigments, binders, coatings and adhesives in paintings. Micro-FTIR played an important role in the detection of anachronistic pigments in the Jackson Pollock affair.[17]

Principle of the Method

IR radiation is invisible and lies on the low energy side of the electromagnetic spectrum, with wavelengths longer than ordinary visible light. The energy associated with this part of the spectrum is not high enough to excite electrons but can cause the bonds in molecules to vibrate. It is as if the atoms in a molecule were connected by springs that can be bent or stretched (i.e. that can vibrate) and that the frequency of the vibrations depends on the strength of the springs and on the masses of the atoms. Molecules therefore will experience a wide variety of vibrational modes depending on the atoms forming them and on the strength of the bonds between these atoms. Since every molecule has a unique structure, the vibrational modes (which are recorded as a series of peaks and valleys)[18] will act as a fingerprint for the molecule.

Micro-Fourier Transform Infrared Spectroscopic Imaging

The development of FTIR — a major technological breakthrough — allowed for the simultaneous measurement of all of the transmitted IR frequencies through the use of an interferometer (consisting of two highly polished mirrors), the output of which could be converted into a spectrum via a mathematical technique developed by the French mathematician Joseph Fourier. This technique led to a greater resolution in comparison to the dispersive IR instruments as well as to a more rapid collection of data.[19]

A further significant improvement to the technique of FTIR was the incorporation in 1981 (at McCrone Associates) of an optical microscope to the FTIR spectrometer. This led to a totally non-destructive approach in the analysis of paintings by allowing the analysis of extremely small samples (as low as a pictogram or less).

Techniques Reliant on Mass Spectrometry

Introduction

Modern materials used by contemporary artists, such as organic pigments, synthetic binders and additives, have been considered challenging to analyse and identify and have only recently been subjected to the same close investigation as the more traditional ones. The difficulty lies in their great number and variety as compared to inorganic compounds, their relatively small amounts and their being sometimes subject to chemical alteration.[20]

Such challenges have been overcome by the relatively new technique of Py-GC-MS and reliant on mass spectrometry that was developed by various laboratories.

The techniques of laser ablation inductively coupled plasma-mass spectrometry (LA-ICP-MS) and accelerator mass spectrometry (AMS) were also identified as very useful tools in the detection of artificially and naturally produced radioactive isotopes which were recently found to play a crucial role in the identification of forged paintings.

Principle of Mass Spectrometry

Mass spectrometry is an important analytical tool that determines the mass of atoms or molecules by measuring the mass-to-charge ratio of

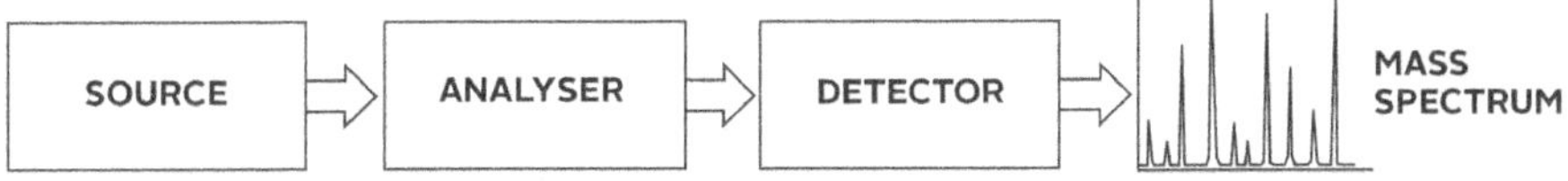

Fig. 60 Typical mass spectrometer

their ions. Once produced, the latter are acted upon by an external electric field and are deflected according to their masses by a magnetic field, the greater deflection being observed for the lighter ions and for those with the highest charge. The results are displayed on a chart referred to as a spectrum from which the atoms or molecules in a sample can be identified through comparison of their determined masses with known masses (see Fig. 60).

Pyrolysis-Gas Chromatography-Mass Spectrometry (Py-GC-MS)[21–23]

Py-GC-MS has proved invaluable in the analysis of synthetic polymers used as binding agents in paints, as well as in the identification of organic pigments. The major problems encountered in identifying synthetic polymers are their high molecular weight, non-volatility and their low solubility in solvents. These problems are addressed by pyrolysis, which allows the breaking down of the polymer into smaller volatile fragments, subsequently separated by gas chromatography and identified by mass spectrometry.

In the gas chromatograph, the components that have been vaporized by pyrolysis are carried by a gas stream (helium or nitrogen) into a tube referred to as the column where they are separated by their interaction with the column's walls. Separation is based on the fact that each component exits the column (i.e. elutes) at a different time known as the retention time. The eluted or removed sample is ionised in the mass spectrograph, followed by the identification of the atoms or molecules (Fig. 61).

High-Performance Liquid Chromatography–Mass Spectrometry (*HPLC-MS*)

The principle of this technique is reminiscent to that of gas chromatography described above. In the case of HPLC-MS, a solvent is used as a

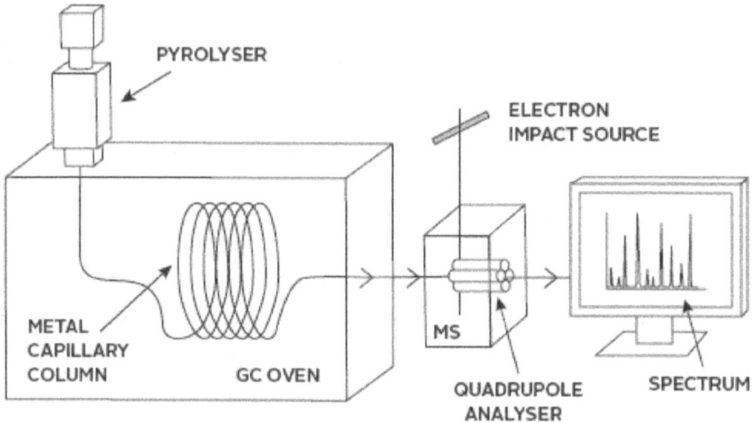

Fig. 61 Arrangement for pyrolysis-gas chromatography

carrier and is forced through the column under high pressures, that can reach up to 400 atm. Minute amounts of solid samples are dissolved in the appropriate solvent, and injection is carried out in a highly automated fashion. Here also, the retention time refers to the time it takes for a particular component to travel through the column to appropriate detectors. In some cases, for a precise identification, the material that is emerging from the column is fed directly into a mass spectrometer.[24]

Laser-Ablation-Inductively Coupled Plasma-Mass Spectrometry (LA-ICP-MS)[25]

LA-ICP-MS is based on the elemental analysis of the artist's paints down to parts per billion (ppb) and can be performed on solid samples without any sample preparation. It is an ideal tool for the analysis of radioisotopes.

A laser beam is focused on the surface of the sample and directly converts it into an aerosol of fine particles which are carried by helium and ionised in a plasma torch.[26] These ions are then analysed in a mass spectrometer, with results obtained in a few seconds (Fig. 62).[27]

Determination of the Radioactive Isotopes ^{137}Cs and ^{90}Sr by LA-ICP-MS

The numerous nuclear tests carried out both during World War II and the Cold War, in particular those of 1955, resulted in significant

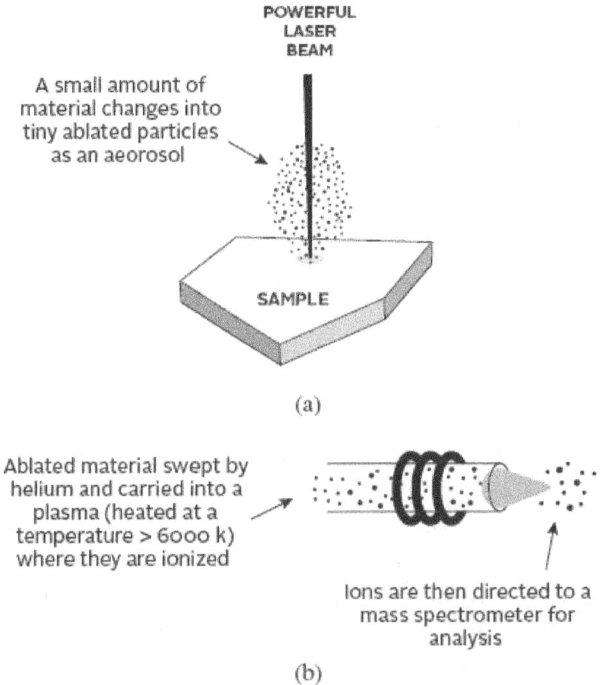

Fig. 62 Principle of LA-ICP-MS (the sequential steps are (a) and (b))

amounts of nuclear fallout. Indeed, the July 1945 nuclear bomb testing in New Mexico, followed by the August bombing of Hiroshima and Nagasaki, resulted in the formation of the radioactive isotopes cesium 137 (^{137}Cs) and strontium 90 (^{90}Sr), which are not naturally found on earth.

Elena Basner, former curator at the Museum of Fine Art in St Petersburg and subsequently consultant of Russian avant-garde art at Bukowski's auction house, thought of using such isotopes to identify forged paintings allegedly created before 1945. Her idea being that the isotopes (^{137}Cs) and (^{90}Sr) would have made their way into the soil and the water throughout the world ending up in all post-1945 art via the natural oils extracted from plants which are used as binding agents in paints.[28,29]

Geochemist Andrey Krusanov and several other researchers of the Russian Academy of Science in St Petersburg validated Basner's hypothesis by using mass spectrometry to identify traces of these two isotopes in paintings created post 1945.[26] It was, however, pointed out that while the detection of such isotopes would certainly reveal a forgery, their absence would not necessarily mean that the painting is authentic.[27] Basner deemed it an ideal way to identify the forgeries of Russian avant-garde works (dated from 1900 to 1930) that flooded the market in the 1980s and 1990s.

Authentication of Paintings by Lead Isotopic Ratio Determination Using LA-ICP-MS

Lead white (basic lead carbonate) has been extensively used in the past as a white pigment in the body of paintings and in their primary layers. It has been a preferred choice for such purposes due to its capacity to dry rapidly when used together with a drying oil and in view of its high durability. However, it was gradually replaced by zinc oxide towards the end of the 18th century, which in turn gave way to titanium white (titanium dioxide) early on in the 20th century.[30]

It was recently found out that the lead isotopic ratio in paintings can be a useful tool in authentication as it can help in identifying the locality of the lead ore from which the metallic lead was extracted before being converted to lead carbonate. The need for a minute sample size (of the order of one microgram) makes it an ideal choice for such an investigation.

The principle of the method is as follows: lead has four natural isotopes which are ^{204}Pb, ^{206}Pb, ^{207}Pb and ^{208}Pb. The last three originate in part from the radioactive decay of ^{238}U, ^{235}U and ^{232}Th, so that their concentration varies in rocks belonging to different locations.

Luckily isotope ^{204}Pb — not resulting from any radioactive decay — is stable, with its concentration not changing with time. Therefore, a plot of the ratios $^{206}Pb/^{204}Pb$ as abscissa (i.e. on the x axis) and of $^{207}Pb/^{204}Pb$ as ordinate (i.e. on the y axis) gives an indication of the origin of the lead ore from which the lead was extracted[31] (Fig. 63).

Accelerator Mass Spectrometry (AMS)

AMS detects naturally occurring long-lived radioisotopes like carbon-14 (^{14}C) and enables the analysis of very tiny samples to estimate the age of a painting canvas or of its wooden frame. To better

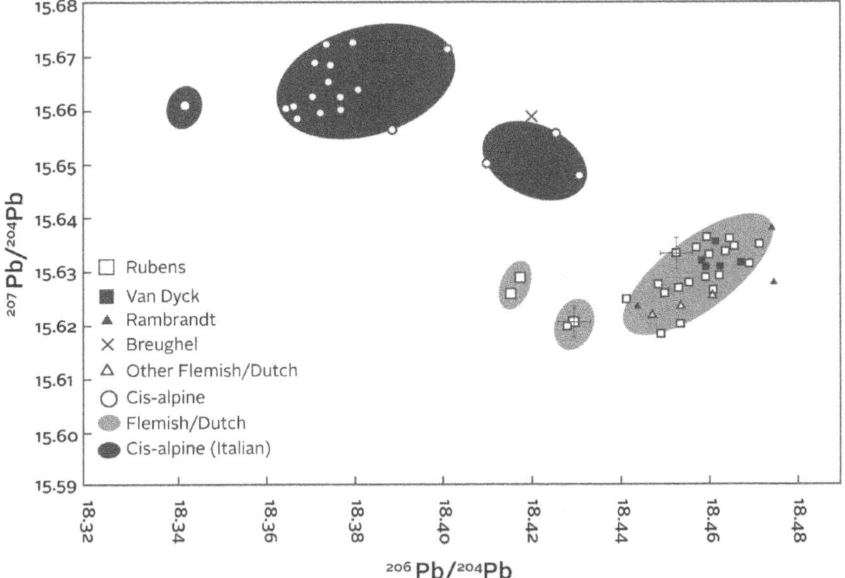

Fig. 63 Origin of the lead ore from which lead was extracted

understand this technique, an introduction to the principles of conventional radiocarbon dating will be provided followed by a description of AMS and its advantages.

Radiocarbon Dating

Radioactive ^{14}C is formed continually in the upper atmosphere by the effect of cosmic rays on nitrogen-14 (^{14}N) atoms. It then combines with oxygen, becoming (radiocarbon) carbon dioxide $^{14}CO_2$ indistinguishable from $^{12}CO_2$ with which it mixes. It is then distributed throughout the atmosphere.

As a result of photosynthesis, the radiocarbon enters the food chain through plants and is absorbed by humans and animals through consumption (Fig. 64). The ratio of $^{14}C/^{12}C$ is stable throughout the lifetime of living organisms, however when an organism dies, the ^{14}C starts decaying at a constant rate without being replenished, and the standard ratio of $^{14}C/^{12}C$ at death is changed because of such a decay.

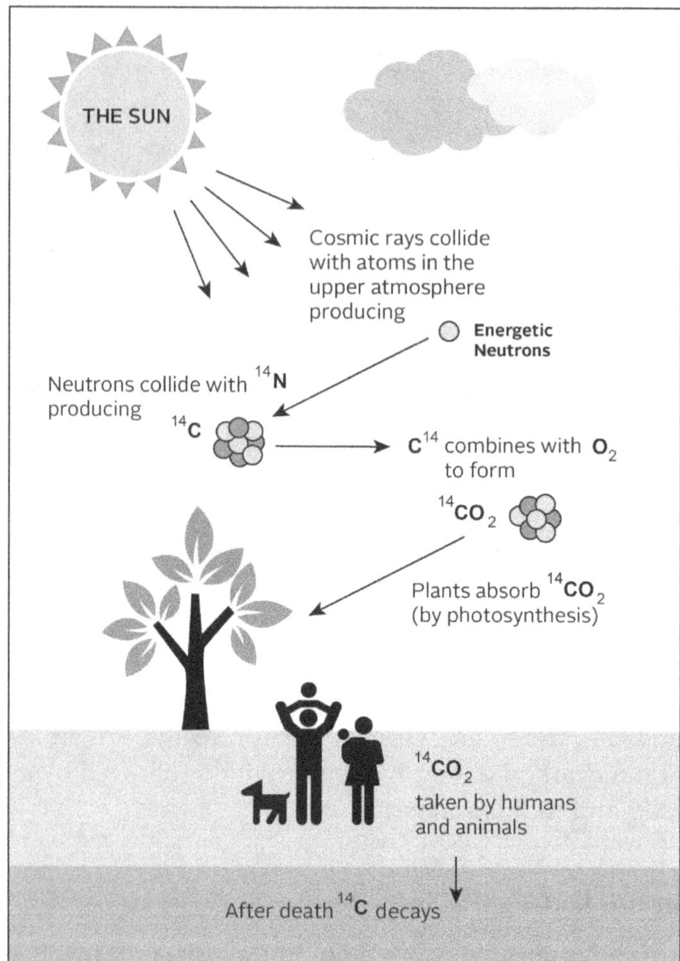

Fig. 64 Formation and decay of ^{14}C

By comparing the ratio of $^{14}C/^{12}C$ in modern samples to that in the dead organism, it is possible to determine the age of the sample.[32,33] Radiocarbon dating can be applied to any organic material.

Advantages of AMS[34]

Whereas 10 g of a sample are needed in standard radiocarbon dating of wood or charcoal and 100 g for dating bones, in AMS 1-2 mg are

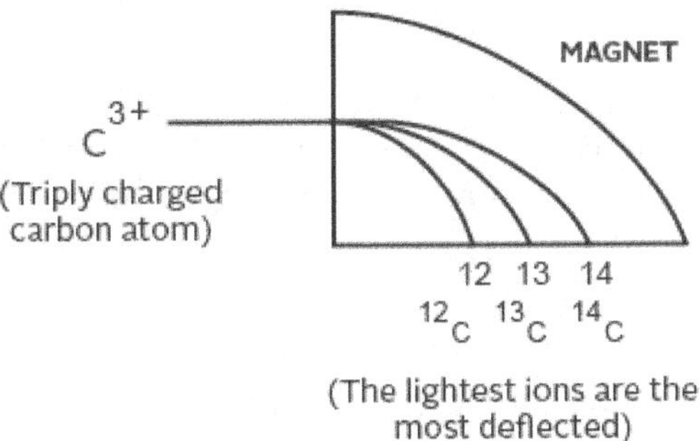

Fig. 65 Principle of mass spectrometry of carbon isotopes

sufficient. Samples are initially converted into a solid graphite form, which by a series of steps, is transformed to a triply charged carbon ion and accelerated to extraordinary high kinetic energies into the mass spectrometer (Fig. 65). The latter has been fine-tuned so that the ^{14}C can be detected without any interference of the more abundant isotope ^{12}C. This technique also has the added advantage that it is a rapid process, only necessitating a runtime of a few hours as compared to one or two days in the case of conventional radiometric techniques.[34]

Some X-Ray Based Techniques

Today, X-ray technology has seen a significant development with a great versatility of approaches for the analysis of artworks. In what follows, some of the many available techniques are introduced together with a description of the principle on which they are based.

X-Ray Fluorescence (XRF)

X-rays are in essence electromagnetic radiation, similar to visible light, but possessing much shorter wavelengths and a far greater penetrating power.

When a primary X-ray strikes a surface with sufficient energy, it can eject electrons from inner shells of the atoms and create vacancies. As a result, the atom becomes unstable and electrons from outer shells will fill these vacancies. In so doing, secondary X-rays are emitted at a unique set of energies and wavelengths that are characteristic of each element (see Fig. 66). This process is referred to as XRF, and it allows the determination of the elemental composition of any given sample. The results are depicted on a spectrum as a number of peaks of various intensities, each occurring at a characteristic energy (wavelength) (see Fig. 67).

Portable XRF spectrometers are now commonly used in museums and galleries, allowing the determination in situ, and in a nondestructive manner, of the elemental composition of a given painting.[35]

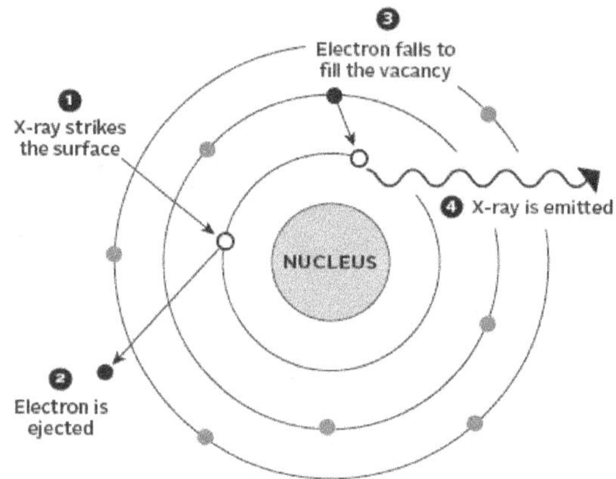

Fig. 66 Principle of XRF

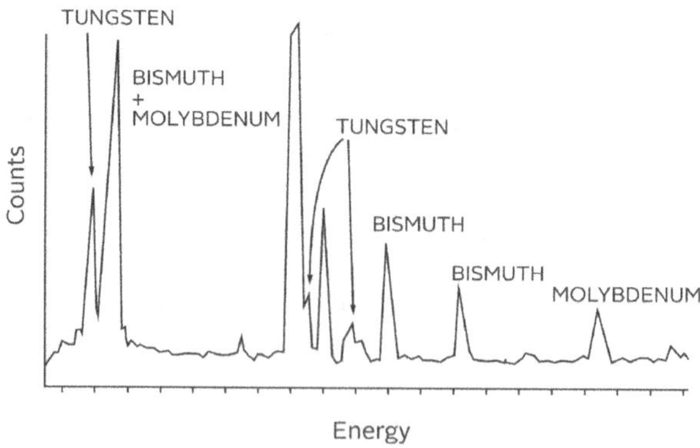

Fig. 67 XRF spectrum

Another approach to XRF is energy dispersive X-ray fluorescence (EDXRF) in which all of the elements present in a sample are concurrently detected. A spectrometer incorporates a special type of detector, connected to a multichannel analyser, allowing all the emitted fluorescent radiation to be collected simultaneously and the energies that characterise each element separated.

Another type of approach to XRF, involving the use of synchrotron radiation as the X-ray radiation will be addressed later in this chapter.

Proton-Induced X-Ray Emission (PIXE)

PIXE is a spectrographic technique used in determining the elemental composition of a sample and is valuable in the non-destructive identification of the composition of pigments in a painting.

This technique is very reminiscent to that of XRF mentioned earlier, the main difference being that instead of primary X-rays, it is energetic protons that excite and liberate the inner shell electrons of atoms creating vacancies. Because of a lower noise to signal ratio, it allows the identification of a much wider spectrum of elements.

X-Ray Radiography (XRR)

Today, radiographic investigation using X-rays is used extensively for the analysis of works of art and is of particular value for examining paintings on wood panels.

X-rays pass easily through layers of paint, but dense pigments such as lead white attenuate them, resulting in bright areas on the XRRs. These provide a wealth of information with regard to the execution technique and to the way the artist conceived and possibly modified the composition of the painting in the underdrawings.

Conventional radiographic analysis uses an X-ray-sensitive film to record, store and display the images. The initial procedure for such a technique was to hold the painting vertically on an easel and use a fluorescent screen to view the image.[36] This arrangement has, in many laboratories, been replaced by placing the paintings horizontally on a lead-lined table with the X-ray source beneath that table, as shown in Fig. 68.[37]

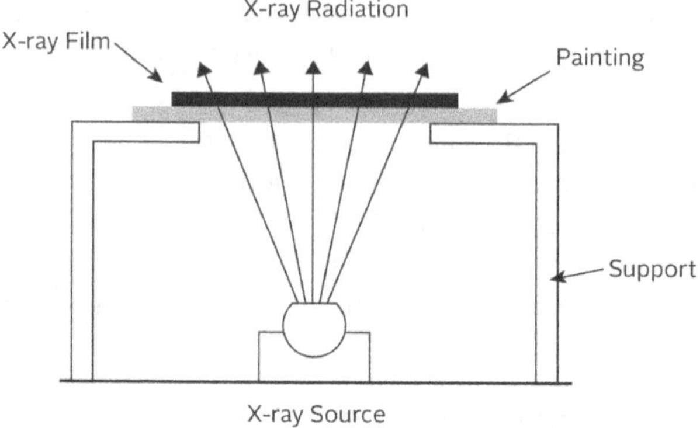

Fig. 68 Arrangement for radiographic analysis

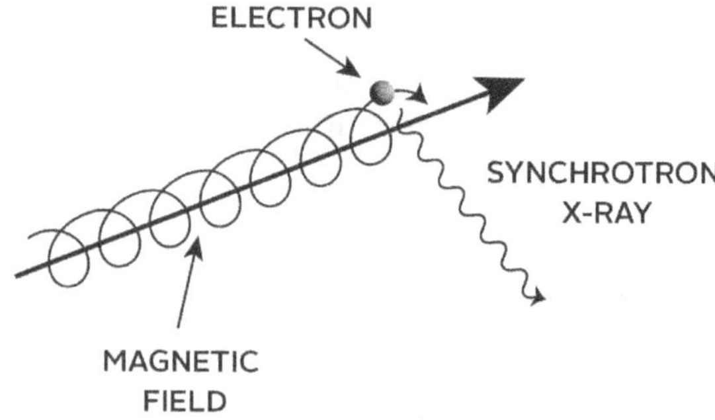

Fig. 69 Production of synchrotron radiation

X-Ray Fluorescence Recorded Using Synchrotron Radiation

A synchrotron is a particle accelerator that emits electromagnetic radiation when charged particles (like fast moving electrons) are accelerated to extremely high velocities. This can be achieved by the appropriate use of so-called bending magnets (see Fig. 69). The emitted radiation can extend from the IR to hard X-ray ranges. Today, with the great advances in accelerator physics, very bright and stable

X-ray beams are produced by synchrotrons and have a wide variety of applications.

Synchrotron X-rays were first observed in 1947, but it was not until the late 1970s that scientists began to produce extremely bright X-rays. In the early 1990s, the first centralized European Synchrotron Radiation Facility (ESRF) was built, followed by the introduction of numerous synchrotron facilities all over the world.

Synchrotron-produced X-ray images of underdrawings in paintings have a much greater resolution than images produced by conventional X-rays tubes, revealing very fine details in the paintings.

The X-ray synchrotron-induced fluorescence technique harnesses the spectrally pure X-ray beam from a synchrotron and causes an XRF, which is characteristic of the chemical elements. It therefore allows the complete identification of elements in an underdrawing and assists in the identification of the pigments used in a given artwork.[38]

X-Ray Diffraction: Principle of the Method

X-ray diffraction (XRD) is a useful technique for the identification of a crystalline material. X-rays generated from an X-ray tube are directed towards a sample and interact with its atoms.

When the conditions of a particular law, known as Bragg's Law are satisfied, a diffraction pattern is obtained, which acts as a fingerprint for the material. In diffraction, the peaks in the resulting X-rays reveal the distances between various atomic planes. XRD is, therefore, a powerful method of characterizing a given material (Fig. 70).

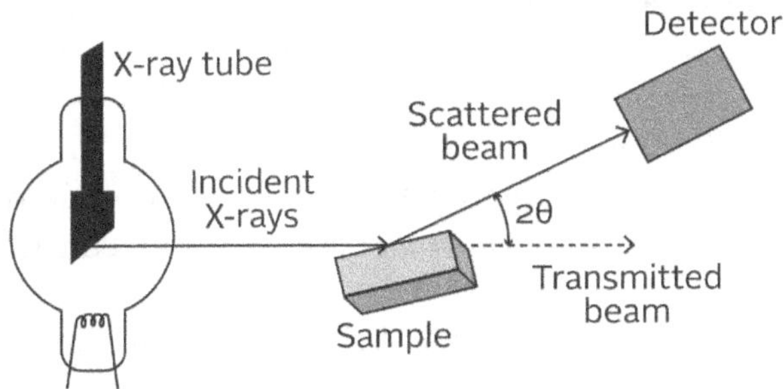

Fig. 70 X-ray diffractometer

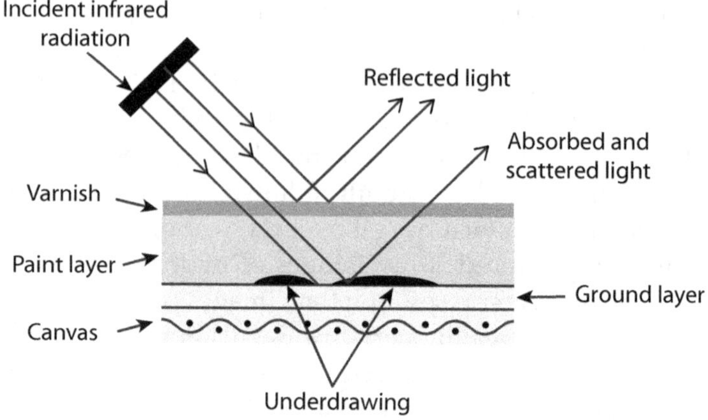

Fig. 71 Principle of IR reflectography

Infrared Reflectography

Principle and Recent Developments

The detection of underdrawings by IR reflectography relies on the fact that IR light penetrates many paint layers further than visible light where it might be absorbed by elements in the background. Light is then reflected back into a sensitive, specially designed camera which can reveal details of the background and produce an image (Fig. 71). When used as sketching materials, graphite or charcoal absorb infrared light very well and are particularly suited for detection by IR reflectography.

The standard approach to IR reflectography is to obtain a complete image of the underdrawing by taking photographs of small regions in the painting, then joining them together piece by piece to form a mosaic-like reflectogram. In the recent past, cameras used for such a purpose generally incorporated a lead sulphide detector, but these produced low resolution and required a long consumption time to manually join the pictures.

The same technique is used today, but with modern instruments that consist of digital cameras with incorporated CCD sensors. A successful attempt to improve the sensitivity of the camera and to considerably decrease the image acquisition time was made through the incorporation of an indium gallium arsenide (InGaAs) array sensor.[39]

Digital Techniques in Art Authentication

Principles of Weave Analysis

In weaving a canvas, the vertical threads mounted on a loom are referred to as the *warp* whereas the horizontal threads constitute the *weft*. In the warp, the threads usually have a steady alignment of almost uniform spacing. The weft threads are woven through interlaced warp threads and show a greater display of variability, because they are individually compacted after each weaving pass (Fig. 72).

Until recently, the standard technique in weave analysis was to extract, in a non-destructive manner, a sample of canvas by selecting a safe location in the painting such as the tacking margin, cutting a small square, then manually counting the number of threads by means of a ruler in both horizontal and vertical directions.

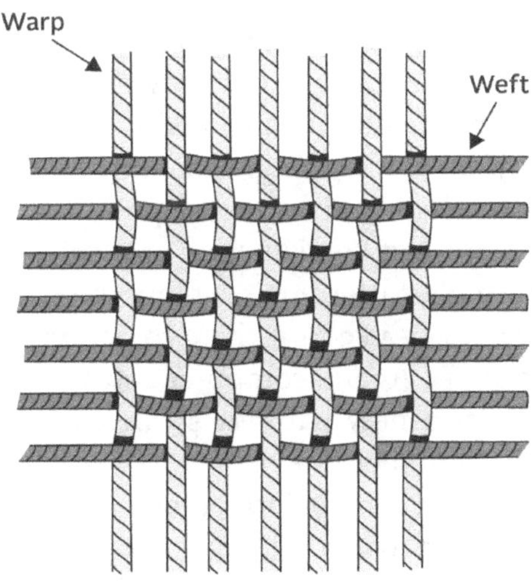

Fig. 72 Weaving in a canvas

In addition to measuring thread counts, canvases were also analysed for fibre types (hemp, linen, machine-woven fine linen, cotton, etc.) with the weight of canvas determined per square centimetre. Since tacking margins may not be present as the canvas could have been lined by gluing a supporting canvas, a more modern approach is to use X-radiographs of the painting. In most cases, these reveal the canvas weave and individual threads. The X-ray films are then mounted on a light box and a ruler is used to count threads. With either approach, a complication arises from the fact that the horizontal threads are not totally parallel to the x axis, as they frequently deviate slightly from a straight line. Furthermore, this mode of thread counting is somewhat inaccurate since selected samples are not necessarily characteristic of the whole canvas. As a result, such an approach cannot play a significant role in authenticity matters.

Recently, a technique has been introduced by Professors C. Richard Johnson, Jr. and Don. H. Johnson from Cornell and Rice Universities respectively. They have developed computer algorithms for the automatic counting of canvas threads from digitized X-ray images of paintings. This method allows the *mapping* of the thread density of both the vertical as well as the horizontal threads everywhere across the painting. These maps reveal a nearly constant thread count in the thread direction, but a random count variation in the other direction. A good clue to identifying the direction of the warp is to select the thread count with the narrower distribution. The automatic thread-counting technique also takes into account the changes in thread angle by producing angle maps that measure the thread angles, which reveal weak and strong cusping.[40]

These weave maps can be visualised as a colour-coded pattern of stripes. If one painting's pattern matches another painting's pattern of stripes, then the canvases used for the two paintings originated from the same bolt of canvas. Again, computer software performs searches to determine whether a new painting's canvas has a thread density pattern that matches another.

As an illustration, if two paintings one meter in width are produced from a two-meters wide roll that is cut in half with one painting arising from the left half and the other from the right, the analysed canvases should show the same variation in *weft*. On the other hand, if two paintings arise from a roll cut from above and below, matching canvases share the same *warp*.[41]

Multispectral Imaging and the L.A.M. Technique (Layer Amplification Method)

The idea of multispectral imaging came as an outgrowth of colour infrared imaging (CRI) which was successfully applied in the 1960s to distinguish artificial camouflage from natural vegetation in wartime (as in the case of the Vietnam war). Subsequently, devices were sought that could extend to regions in the electromagnetic spectrum other than the infrared, distinguishing between blue, green, red and near IR.[42]

This was achieved in the 1970s with the development of a new sensor (the four-band multispectral scanner MSS), which could make digital images in these four spectral regions. This scanner was used in the first Landsat satellite program of the US named "Earth Resources Technology Satellite 1", for taking radiometric images of Earth. Further technological progress led to multispectral imaging covering the UV to the IR regions of the electromagnetic spectrum.[43]

As a tool for artworks, however, these earlier multispectral imaging systems were problematic given the high level of light required to properly illuminate a canvas (about 130,000 watts of power for a 12 m^2 canvas).[44]

Pascal Cotte and Francois Dupuy from Lumiere Technology resolved this problem through the invention of the first multispectral high-definition camera, with a lighting system which required only 2400 watts of power, without compromising the illumination's spatial homogeneity.

This led to a new analysis technique, the layer amplification method (L.A.M.), which allowed the inside of the paint layers to be seen one by one as if in a peeled onion, and their composition determined. The integrity of the artwork was preserved, and the pigment materials were identified at each pixel thanks to the method pioneered by Professor Mady Elias, head of the Optics and Art Group at the Paris Institute of Nanosciences (INSP) — the latter is a mixed research unit of the French National Research Center (CNRS) and the University of Paris. This technique also gave IR reflectographs of exceptionally high sharpness at three different depths below the pictorial surface, allowing the construction of a map of apparent restorations in the painting, all of crucial importance in authenticity work.

How Does Cotte's Multispectral Imaging Work?

An average digital camera records around 20 million pixels at three different wavelengths of the electromagnetic spectrum, corresponding to three different colours. Lumiere's camera records 240 million pixels at 13 different wavelengths, four of which are outside the visible spectrum.

The 13 different filters are supported in a specially designed filter barrel (Fig. 73). Each filter allows the sensor to see only a narrow bandwidth of the light spectrum. The sensor moves by a stepper motor on 20,000 vertical lines and scans the whole surface with a first filter then a second, etc. This translates as a beam of light sweeping the image 13 times, without any risk for the painting.[45,46]

A complicated series of graphs and numbers are obtained (around 24 gigabytes of data), which, when fed into a computer using a special software designed by Cotte, allows the eventual generation of a

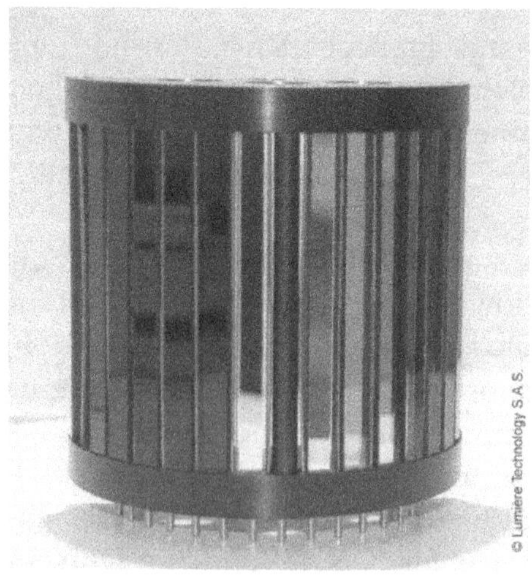

Fig. 73 The multispectral camera's 13-filter wheel

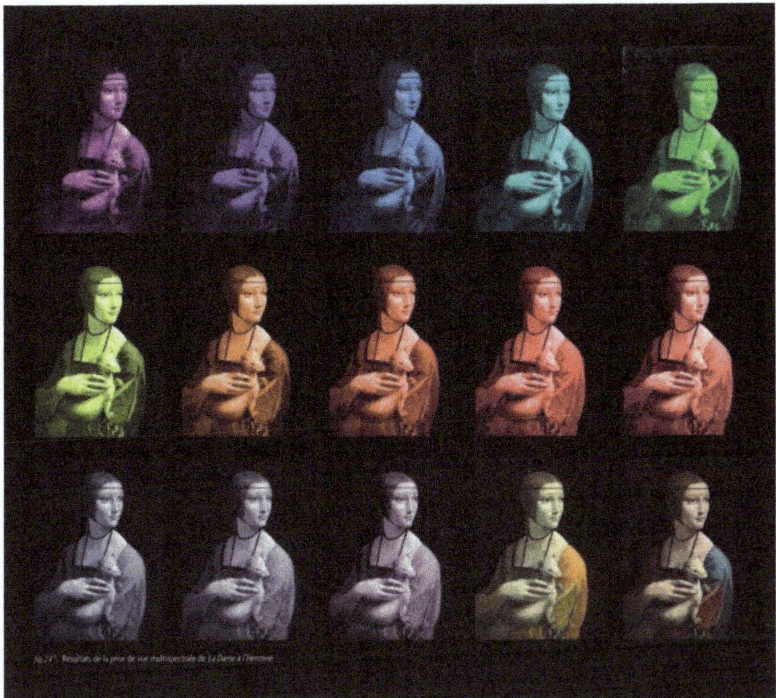

Fig. 74 Multispectral views in colours of *The Lady with an Ermine* by Leonardo da Vinci

number of photographs ranging from the UV to the IR ranges of the electromagnetic spectrum (Fig. 74). The true colours of the artwork as perceived by an observer are obtained through the superposition of these photographs after a virtual removal of the varnish revealing, layer by layer, the story and evolution of the picture.

Dendrochronology

Dendrochronology, developed in the first half of the 20th century, is used to determine the age of wooden objects and can be applied to date panels or frames in paintings.

At the beginning of the year, a thin walled cellular structure develops in a tree, it changes at the end of the year creating visible dark and lighter areas or "growth rings" sometimes also referred to

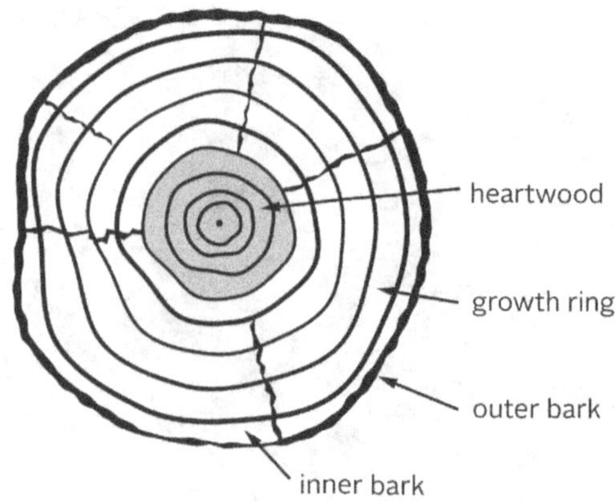

Fig. 75 Tree rings

as annual rings. These rings are generally distinguishable from one another, they form concentric bands around the pith of the tree and can be clearly observed in a horizontal cross-section cut through the trunk of the tree (Fig. 75). It is by counting the number of rings that one can establish the age of the tree.

The method is only applicable to temperate zones where trees generally make one growth ring per annum — in tropical climates, it is difficult to discern the material laid down annually. Dendrochronology confirms the felling date of the tree from which the wooden object was cut.[47–50]

The relative width of the rings depends on growing conditions: better growing conditions result in wider rings, while poorer conditions produce narrow rings. Given that these widths are highly dependent on the climate in which the tree grows, for a viable dating of a wooden object it is crucial to identify the place where the tree grew.

Trees belonging to the same region will produce similar patterns of ring widths, and it is therefore possible to create databases and chronologies for various regions in the world.

Measurement of the width of the ring can be carried out manually by means of a magnifying glass and a scale or can be made through computer tomography, where the data is fed as a digital image.[51,52] Either will give generally reliable dating from 50 to 100 rings. Using an appropriate computer program, the sample tree ring sequence is then compared to the tree ring chronology of the same region, and thus the age of the wooden object is established.[51]

A great deal of work was done on oak, but the technique is also applicable to other types of wood such as pine, beech, spruce, fir and larch.[51,53]

Epilogue

What we call the beginning is often the end
And to make an end is to make a beginning.
The end is where we start from.

T. S. Eliot[1]

Navigating through the art world today, one finds it riddled with complexity; market prices have soared — not least because of the growing tendency for artworks to be viewed as commodities, desired by some for the social status they impart, and by others, as an attractive venue investment for mitigating tax-liabilities. The upward spiral is accompanied by further incentives to forge, and the highly vulnerable art market is flooded with fakes and repeatedly exposed to outright fraud. The relentless quest to detect and curb forgeries faces new challenges as clever forgers find ways to outwit the system and unscrupulous forensic experts further muddy the waters. Sceptical scholars and authenticators are reluctant to voice any opinion for fear of libel. Art critic Walter Robinson perfectly captures the essence of the conundrum:

Since the entire market is entirely irrational, it can't be rationally interpreted[2]

As key players try to untangle this very tight web, a competent use of science has come to play a crucial role. Given how scientific revelations challenge the trained eyes of art connoisseurs, however, scientific verdicts have been viewed with suspicion. While collectors and museums were cautiously in support of the contribution of science to authenticity matters, auction houses in particular tended to be more dismissive and did not regard sophisticated research to be within their purview.

The use of science in unlocking the Beltracchi and Knoedler affairs, as well as the more recent Old Masters scandal, has led to somewhat greater trust. The detection of anachronistic elements in paintings, in these cases using relevant scientific techniques, has highlighted the importance of competent science in confirming or challenging the connoisseur's evaluation.

Even more telling in this regard is the recent purchase by Sotheby's of the reputable scientific firm Orion Analytical, strongly suggesting that scientific analysis has become part of the auction house's due diligence exercise.[3–7]

Had scientific evidence been sought in the case of van Meegeren's forgeries, it would have revealed that in his painting *The Adulteress* he used the anachronistic cobalt blue pigment and that the crackle in the underground layer did not match the one at the surface as later indicated by XRR.[8] This would have avoided a great deal of subsequent damage to the aesthetic understanding of Vermeer's oeuvre.

When trained technical art scientists, working at the laboratories department of the Belgium Royal Institute of Cultural Heritage investigated an abstract painting attributed to the Russian avant-garde artist Vasily Kandisky, they could determine the relevant tools needed for their analysis. By using micro-Raman spectroscopy, and with the aid of an internally generated Raman reference library which included the spectra of 300 synthetic organic pigments, they could identify certain pigments resulting in their unequivocal conclusion that the painting was a forgery.[9]

The need for an appropriate scientific technique is nowhere better exemplified than in the case of Wolfgang Beltracchi, where the anachronistic titanium dioxide in a Campendonk painting was detected in an underlayer through the use of EDXRF.[10] Only a well-trained art-scientist could have appreciated the importance of analysing all the layers in the artwork, being conscious of the fact that material present on the surface may appear unsuspicious and lead to incorrect conclusions.

In the case of the recent Old Masters scandals, after careful scrutiny and approval by prominent connoisseurs, the painting representing *Saint Jerome* was attributed to the circle of the 16th-century artist Parmigianino. However, James Martin of Orion Analytical later identified the anachronistic pigment phthalocyanine green in samples taken from this artwork. Such a pigment could only have been detected by a suitable tool such as micro-FTIR or micro-Raman and not by XRF, a technique routinely used in many laboratories for the identification of pigments (Fig. 76).[10]

In spite of all the successful uncovering of forgeries by scientific means, there are those who are still quite sceptical about scientific verdicts even when they arise from trustworthy laboratories.

We may recall the Jackson Pollock affair, where, convinced that the 32 paintings he discovered in his father's attic were indeed authentic, Alex Matter responded with some suspicion to the verdict of inauthenticity given on three of these paintings by the Harvard University Center for the Technical Study of Modern Art.[11]

A public relations firm appointed by Matter made the following statement:

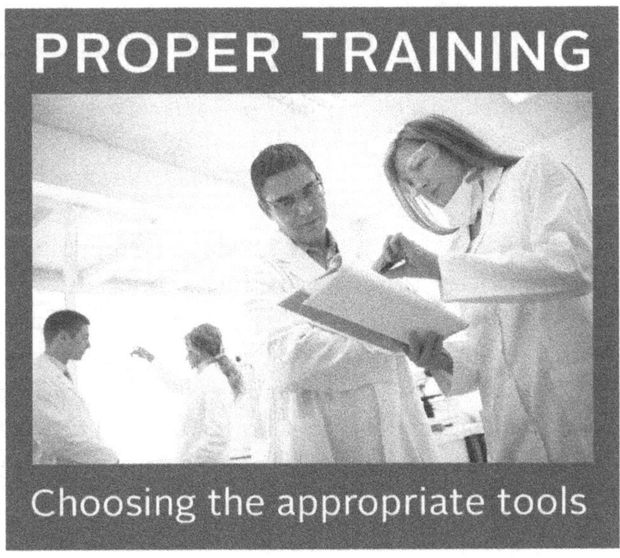

Fig. 76 The proper approach to authentication

Many attribute them to Jackson Pollock and nothing in the Harvard report effectively challenges that. The authentication of works of art is still more art than science.[11]

Matter was unconvinced by the Harvard verdict until James Martin, from Orion Analytical, mass-studied a number of these alleged paintings and confirmed the Harvard results of inauthenticity.[12]

In the aforementioned Old Masters scandal, Mr Weiss, who provided the forged Hals to Sotheby's, strongly contested the claim that it was a forgery, and mistrusting the results provided by James Martin that revealed anachronistic materials, requested that further tests be carried out. He would declare in an interview to the *Financial Times* that he was "yet to be convinced" that it is a forgery.[13]

There is no doubt that greater trust in the verdicts passed by qualified laboratories is still very much in need.

Unfortunately, the art world is also witnessing the emergence of some self-proclaimed art–scientists who possess questionable ethical codes giving ill-informed and misleading so-called expert opinions.

Such "charlatanism"[14] raises grave concerns and has been clearly described by Tom Flynn:

With the art market booming again, and fakes proliferating, it now seems clear that charlatanism is not confined to the faking of paintings and the forging of provenance documentation. It also extends to a new breed of self-appointed and ethically compromised 'forensic scientists' willing to issue certificates of authenticity for inauthentic works in return for financial kickbacks.[15]

To address such a problem, there needs to be a greater number of specialised laboratories recognised on the national/international levels to assure collectors they are dealing with genuine artworks. Technical skills that address problems of authentication in art need to be developed through properly structured training and educational programmes. At present, such laboratories are largely the realm of some specialised museums and universities and of a precious few consultants worldwide.

Tangible efforts are being made in that direction by the *Authentication in Art Foundation*, which calls for educational reforms in art history and conservation and for the establishment of a unified standard within technical research in art.[16] The Foundation recognises the need for developing strategies for fostering greater interaction between connoisseurs and conservators as well as an integrated analytical and investigative methodology that would meet the evidentiary requirements of the courts.[17]

As stipulated in the guidelines on education presented by the Foundation, art history and conservation curricula need to focus more on the concept of *authenticity*, as well as on the role that scientific protocols play within the legal system, stressing the importance of interdisciplinary collaboration.[16]

Universities should be encouraged to develop graduate Art Authentication programmes for students aiming to work as art scientists, the likes of those presently offered at the University of Melbourne[18] and at the University of Delaware, which offer curricula, websites, and workshops promoting Technical Art History and educating a variety of audiences.[19] Such activities would ensure that minimum standards are maintained in the long run and that safeguards are instated for greater accountability in the processes of technical art authentication.

The exchange of data and the sharing of findings in works of art is also crucially important. Through its Science department, The National Gallery, London has investigated and recorded all the materials characteristic of paintings in its collection for the period extending from 1260 AD to the early 20th century. This proved to be of great value in the investigation of artworks and led to the uncovering of a number of forgeries which were shared with the public in the National Gallery's exhibit of June 2010 entitled *Close Examination: Fakes, Mistakes and Discoveries*.[20]

When the unsigned painting *Portrait of Two Children* from the Hispanic Society of America was claimed to have been created by the famous Valencian painter Joaquín Sorolla (1863–1923), EDXRF studies of this painting indicated an unusually high proportion of zinc as compared to data obtained on 50 of this artist's authenticated paintings, suggesting that the painting was a forgery.[21]

In a similar approach, Joanna E. Russell et al., using a wide spectrum of up-to-date techniques, investigated the materials and techniques used in 38 paintings attributed to Francis Bacon (1909–1992). This project involving the British Museum and an academic institution emphasised the importance of such data which has already proved to be of great value in the authentication of paintings attributed to Francis Bacon.[22]

All these collected data are examples of information which should, if not already available, be made accessible to arttechnical experts. As noted by Heike Stege: "a broader exchange of data, findings in art works and the widespread meta-information among experts would help to use this rich pool of evidence for authenticity questions more efficiently."[23]

Science is not only good at *falsification* but can also play an important role in *authentication* when made part of *an integrative approach* that also brings together art historians, connoisseurs, curators, etc.

Duane Chartier and Fred Notehelfer have called for such an interdisciplinary approach writing, "there is great need for the integration of a scientific approach to the authentication of works of art. What is required is a task that is often paid lip service to but rarely performed on a day to day basis — true interdisciplinary work."[24]

When connoisseurs were unable to authenticate van Gogh's paintings *Sunset at Montmajour* and *Still Life with Meadow Flowers and Roses,* on the grounds of style and aesthetic value alone, it was the successful convergence of art historical evidence and scientific investigation that led to the final verdict of authenticity. The crucial factors here were provided on the one hand by van Gogh's letters to his brother Theo and on the other hand by the scientific investigations revealing the underdrawing in the case of van Gogh's *Still Life* and the close analysis of the canvas and the palette in the case of the sunset painting. All the confirmatory scientific tests subsequently carried out on these two artworks were crucially important but would have never been considered as sufficient to authenticate these paintings without the art–historical evidence.

Again, the importance of the need for close collaboration between honest connoisseurship and competent science (Fig. 77) is well illustrated in the case of the authenticated *Infanta Margarita*[25] attributed to Edouard Manet, but lacking provenance and a signature. While understanding the initial reluctance of connoisseurs unequivocally to attribute the painting to Manet, despite general agreement that the style was that of the painter, Albert Boime and Alexander Kossolapov, wrote,

> *Regrettably, many art experts still mistrust the methods of the scientist, but it is our firm belief that not only is this gulf not unbridgeable, in future this*

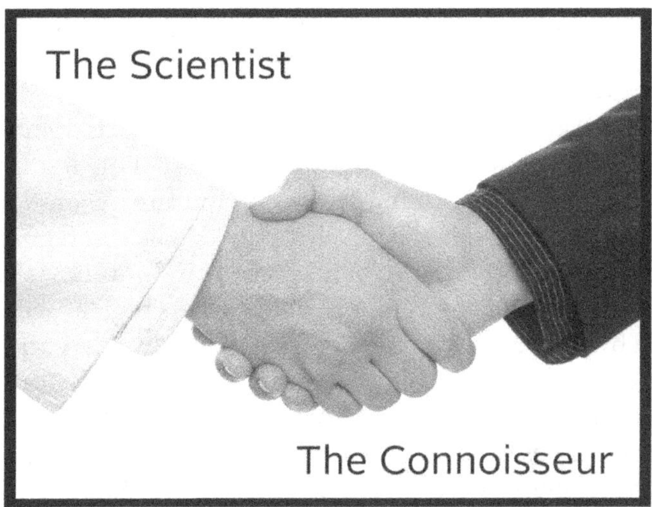

Fig. 77 An integrated approach

collaboration will be the norm. It has always been a given that in a situation where provenance and signature are absent, the convergence of agreement on the part of both art expert as to style and physical appearance (what has been called the work's internal evidence) and conservation scientist as to technical and physical properties (the external evidence) establishes authenticity...[25]

Those were prescient words, as revealed by the recent Old Masters scandal, calling into question the long-held belief that connoisseurs can recognise the hand of a master by closely observing and inspecting the artwork. Several prominent art historians had expressed their enthusiasm and applauded the Old Masters paintings which were proven through scientific examination to be forgeries. The need for true interdisciplinary work can certainly no longer be given lip service, and there cannot be exclusive reliance on the judgement of connoisseurs on authenticity without the backing of scientific evidence.

Laws protecting authenticators from irresponsible lawsuits that threaten the natural functioning of the art market could add trust and protect gallery dealers, authors of catalogues raisonnés, auction houses and experts who provide their candid opinions on artworks.

As was mentioned earlier, the relentless quest to guard against forgeries, and address many of the problems faced by today's imperfect art world has led to the two important initiatives: synthetic DNA acting as a marker for the artist's artwork, and the adaptation of the blockchain technology to artworks. These would allow living artists to ascertain their own artwork for present and future generations of collectors with the blockchain acting as a public ledger, safeguarding the art market from manipulation for all present and future stakeholders.

Let us dream that the day will come when buyers will be more influenced by the intrinsic quality of the art than by the work's attribution to a *recognised* artist, charlatanism is controlled, forgeries are greatly reduced and the art world, not anymore ridden by complicity in the face of unbeatable self-interest, is no longer rocked by scandal.

Edward Helmore's words deserve some reflection:

> *Problems in the market tend to occur when we forget why art is created in the first place. When money gets turned into art that is one thing, but when art gets translated into money, it's a problematic transaction.*[26]

References

INTRODUCTION

1. Helmore, E. (2016). 'Art market in 'mania phase' and risks bursting of the bubble, report says'. *The Guardian*. Available at: https://www.theguardian.com/artanddesign/2016/jan/17/art-market-mania-phase-bubble-report [accessed 30 January 2017].

PART I

ATTRIBUTION

1. Ernst, R. P. (2014). 'Science and art — My two passions', in Sgamellotti, A., Brunetti, B. G. and Milani, C. (eds.), *Science and Art: The Painted Surface*, The Royal Society of Chemistry, London, p. 25.
2. Chartier, D. R. and Notehelfer, F. G. (1988). 'Authentication: Science & art at odds?', in McCrone, W., Chartier, D. R. and Weiss, R. J. (eds.), *Scientific Detection of Fakery in Art*, SPIE, Bellingham, vol. 3315, p. 76.
3. Dutton, D. (1998). 'Forgery and plagiarism', in Chadwick, R. (ed.), *Encyclopedia of Applied Ethics*, Academic Press. Available at: http://www.denisdutton.com/forgery_and_plagiarism.htm [accessed 16 September 2014].
4. Felichenfeldt, W. (2014). 'The fakes controversy — Van Gogh Fakes: The Wacker affair, with an illustrated catalogue of the forgeries'. Available at: http://www.vggallery.com/misc/fakes/wacker.htm [accessed 13 May 2014].
5. Dutton, D. (1993). 'Art hoaxes', in Stein, G. (ed.), *The Encyclopedia of Hoaxes*, Gale Group, Detroit. Available at: http://denisdutton.com/van_meegeren.htm [accessed 16 October 2014].
6. Bredius, A. (1937). 'A new Vermeer: Christ and the disciples at Emmaus', *The Burlington Magazine*, **71**(416), pp. 210–211.

7. Anon (1999). 'Faker who flooded the art world jailed for 6 years'. *The Guardian*. Available at: http://www.theguardian.com/uk/1999/feb/16/5 [accessed 13 May 2014].
8. Levingston, S. (2009). Book review: *Provenance* by Laney Salisbury and Aly Sujo. Available at: http://www.washingtonpost.com/wp-dyn/content/article/2009/07/24/AR2009072401421.html [accessed 13 May 2014].
9. Landesman, P. (1999). 'A 20th-century master scam'. Available at: http://www.nytimes.com/1999/07/18/magazine/a-20th-century-master-scam.html [accessed 2 November 2014].
10. Nicholas, S. (2015). 'Is this Leonardo da Vinci drawing a £100m masterpiece…or a worthless forgery'. Available at: http://www.express.co.uk/news/history/623077/Leonardo-da-Vinci-La-Bella-Principessa [accessed 8 June 2016].
11. Charney, N. (2015). *The Art of Forgery: The Minds, Motives and Methods of the Master Forgers*, Phaidon Press, pp. 181–183.
12. Leigh, D. and Borissova, E. (2004). 'How forgery turned £5,000 painting into £700,000 work of art', *The Guardian*. Available at: http://www.theguardian.com/uk/2004/jul/10/arts.artsnews [accessed 4 November 2013].
13. Princeton University Art Museum (2014). 'Provenance research'. Available at: http://artmuseum.princeton.edu/collections/provenance-research [accessed 20 May 2014].
14. Panofsky, E. (1940). 'The history of art as a humanistic discipline', in Greene, T. M. (ed.), *The Meaning of the Humanities*, Princeton University Press, New Jersey, pp. 89–118.
15. Freidlander, M. J. (1960). 'On intuition and the first impression', *On Art and Connoisseurship*, Beacon Press, Boston, p. 173.
16. Wieseman, M. E. (2010). *A Closer Look: Deceptions and Discoveries*, Yale University Press, National Gallery, London, p. 37.
17. Wolheim, R. (1973). 'Giovanni Morelli and the origins of scientific connoisseurship', *On Art and the Mind: Essays and Lectures*, Allen Lane, London, pp. 209–219.
18. Morelli, G. (1892). *Italian Painters: Critical Studies of Their Works*, John Murray, London, pp. 1–58.
19. Wind, E. (1963). 'Critique of connoisseurship', in *Art and Anarchy*, Northwestern University Press, Illinois, pp. 30–46.
20. Fernie, E. (1995). *Art History and its Methods: A Critical Anthology*, Phaidon Press, London, pp. 103–115.
21. Byrne, D. (2010). 'Connoisseurship today'. Available at: http://artintheblood.typepad.com/art_history_today/2010/07/connoisseurship-today.html [accessed 12 April 2014].

THE FORGER

Sandro Botticelli

1. Dempsey, C. (2016). *Sandro Botticelli: (Grove Art Essentials)*, Oxford University Press, p. 1.
2. Lightbown, R. W. (2014). 'Sandro Botticelli Italian painter'. Available at: https://www.britannica.com/biography/Sandro-Botticelli [accessed 5 February 2017].
3. History of Painters (2017). 'Bonfire of the Vanities'. Available at: http://www.historyofpainters.com/bonfire_vanities.htm [accessed 5 February 2017].
4. Ask Art (2017). 'Umberto Giunti'. Available at: http://www.askart.com/artist/Umberto_Giunti/11103295/Umberto_Giunti.aspx [accessed 28 January 2017].
5. Morris, R. C. (2004). 'Masters of the art of forgery'. *The New York Times*. Available at: http://www.nytimes.com/2004/07/31/style/masters-of-the-art-of-forgery.html?_r=0 [accessed 28 January 2017].
6. Owen, R. (2004). 'Duped art experts praise a master forger'. *The Times*. Available at: http://www.thetimes.co.uk/tto/news/world/article1976850.ece [accessed 28 January 2017].
7. The National Gallery (2017). 'Madonna of the veil'. Available at: https://www.nationalgallery.org.uk/paintings/research/the-madonna-of-the-veil [accessed 3 February 2017].
8. Wullschlager, J. (2010). 'The Dürer that wasn't'. *Financial Times*. Available at: https://www.ft.com/content/0253eafe-8564-11df-aa2e-00144feabdc0 [accessed 28 January 2017].
9. Werth, C. (2010). 'Forged paintings on display — for real'. *Newsweek*. Available at: http://europe.newsweek.com/forged-paintings-display-real-72753 [accessed 28 January 2017].
10. The Art of Copies (2017). 'The Madonna of the veil; Botticelli?' Available at: https://theartofcopies.wikispaces.com/The+Madonna+of+the+Veil%3B+Botticelli,+%3F) [accessed 3 February 2017].

Heinrich Campendonk

1. Deutsche, W. (2017). 'How Beltracchi, the world's most famous art forger, plays with the market'. Available at: http://www.dw.com/en/how-beltracchi-the-worlds-most-famous-art-forger-plays-with-the-market/a-18436266 [accessed 20 February 2017].
2. Paterson, T. (2011). 'Time runs out for Germany's master forgers'. *The Independent*. Available at: http://www.independent.co.uk/arts-entertain-

ment/art/news/time-runs-out-for-germanys-master-forgers-2375970.html [accessed 20 February 2017].

3. Hammer, J. (2012). 'The greatest-fake-art scam in history'. *Vanity Fair*. Available at: http://www.vanityfair.com/culture/2012/10/wolfgang-beltracchi-helene-art-scam [accessed 3 January 2014].
4. Tarsis, I. (2012). 'Forgery and fraud in the art market, as usual'. *Center for Art Law*. Available at: http://itsartlaw.com/2012/01/14/forgery-and-fraud-in-the-art-market-as-usual/ [accessed 4 December 2013].
5. Artkabinett (2011). 'Art forger receives light sentence'. Available at: http://artkabinett.com/ak-file/art-forger-receives-light-sentence [accessed 5 January 2014].
6. Wloszczyna, S. (2015). 'Beltracchi — The Art of the Forgery'. Available at: http://www.rogerebert.com/reviews/beltracchi-the-art-of-the-forgery-2015 [accessed 5 January 2014].
7. Thomas, J. M. (2006). 'Faraday and Franklin', *Proceedings of the American Philosophical Society*, **150**(4), pp. 523–541.
8. Eastaugh, N. (2012). 'Forging Ahead: The Jägers-Beltracchi case and the evolving role of scientific and technical art historical analysis in authentication'. *International Institute for Conservation of Historic and Artistic Works*. Available at: https://www.iiconservation.org/node/3211 [accessed 5 January 2014].
9. Spiegel Online International (2012a). 'Interview online with Wolfgang Beltrachi: Confessions of a genius art forger'. Available at: http://www.spiegel.de/international/germany/spiegel-interview-with-wolfgang-beltracchi-confessions-of-a-genius-art-forger-a-819934.html [accessed 6 January 2014].
10. Spiegel Online International (2012b). 'Interview with Wolfgang Beltrachi: Confessions of a genius art forger. Part 3: I painted because I wanted to'. Available at: http://www.spiegel.de/international/germany/spiegel-interview-with-wolfgang-beltracchi-confessions-of-a-genius-art-forger-a-819934-3.html [accessed 20 February 2017].
11. Meyer, L. (1983). 'Forgery and the anthropology of art', in Dutton, D. (ed.), *The Forger's Art: Forgery and the Philosophy of Art*, University of California Press, Berkeley, p. 88.
12. Switzer, R. (2014). Dean of Undergraduate Studies, The American University in Cairo, e-mail sent to Jehane Ragai, 25 March 2014.

Marc Chagall

1. Alberge, D. (2014). 'The man whose real Chagall could now be burnt as fake'. Available at: https://www.theguardian.com/artanddesign/2014/feb/01/chagall-could-be-furnt-fortune-or-fake [accessed 23 August 2014].

2. Daley, M. (2014). 'Good science; over-reaching science; over-promotedscience'. Available at: http://artwatch.org.uk/good-science-over-reaching-science-over-promoted-science/ [accessed 23 August 2014].
3. BBC News (2014). 'Fake $100,000 Marc Chagall painting to be destroyed'. Available at: http://www.bbc.com/news/entertainment-arts-26000331 [accessed 23 August 2014].
4. Mendick, R. (2014). 'Family fights to save £100,000 'Chagall' from the furnace'. Available at: http://www.telegraph.co.uk/culture/art/art-news/10612313/ [accessed 23 August 2014].

Francesco Francia

1. Vasari, G. 'Francesco Francia (1450–1517). Goldsmith and painter of Bologna'. *Vasari's Lives of the Artists*. Available at: http://members.efn.org/~acd/vite/VasariFrancia.html [accessed 28 January 2016].
2. The National Gallery. 'The Virgin and Child with an Angel'. Available at: https://www.nationalgallery.org.uk/paintings/research/the-virgin-and-child-with-an-angel [accessed 28 January 2016].
3. Wullschlager, J. (2010). 'The Dürer that wasn't'. *Financial Times*. Available at: https://www.ft.com/content/0253eafe-8564-11df-aa2e-00144feabdc0 [accessed 28 January 2017].
4. Switzer, R. (2014). Dean of Undergraduate Studies, The American University in Cairo, e-mail sent to Jehane Ragai, 6 April 2014.

Fernand Léger

1. Herbert, R. L. et al. (2002). *From Millet to Léger: Essays in Social Art History*, Yale University Press, p. 115, ISBN 0300097069.
2. Fam People (2012). 'Fernand Léger: Biography'. Available at: http://www.fampeople.com/cat-fernand-l%C3%A9ger [accessed 2 March 2017].
3. The Art Story (2017). 'Fernand Léger'. Available at: http://www.theartstory.org/artist-leger-fernand.htm [accessed 2 March 2017].
4. Encyclopaedia Britannica (2017). 'Contrast of forms. Work by Léger'. Available at: https://www.britannica.com/topic/Contrast-of-Forms [accessed 2 March 2017].
5. AZ Quotes (2017). 'Fernand Léger quotes'. Available at: http://www.azquotes.com/author/28596-Fernand_Leger [accessed 2 March 2017].
6. Wikipedia (2017). 'Peggy Guggenheim Collection'. Available at: https://en.wikipedia.org/wiki/Peggy_Guggenheim_Collection [accessed 14 March 2017].

7. Peggy Guggenheim Collection (2017). Available at: http://www.guggenheim-venice.it/inglese/museum/peggy.html [accessed 10 March 2017].
8. Gannon, M. (2014). 'Guggenheim painting proven to be a fake'. *Live Science*. Available at: http://www.livescience.com/43178-nuclear-bomb-forged-painting.html [accessed 4 October 2014].
9. Zolfagharifard, E. (2014). 'Léger painting revealed as a fake: Traces of cold war bombs are found on canvas that was meant to have been painted in 1913'. *Mail Online*. Available at: http://www.dailymail.co.uk/sciencetech/article-2556789/Leger-painting-revealed-fake-Traces-Cold-War-BOMBS-canvas-meant-painted-1913.html [accessed 4 October 2014].

Old Masters

1. Luck, A. (2016). ''Moriarty of the Old Master' pulls off the art crime of the century: Market in crisis as experts warn £200m of paintings could be fake'. *Mail Online*. Available at: http://www.dailymail.co.uk/news/article-3817580/Moriarty-Old-Master-pulls-art-crime-century-Market-crisis-experts-warn-200m-paintings-fakes.html [accessed 17 February 2017].
2. Noce, V. (2016). 'French police seize painting attributed to Cranach, owned by the Prince of Liechtenstein'. *The Art Newspaper*. Available at: http://old.theartnewspaper.com/news/news/french-police-seize-painting-attributed-to-cranach/ [accessed 17 February 2017].
3. Blumenfeld, C. (2016). 'Exclusif: après l'affaire Cranach, l'affaire Gentileschi à la National Gallery'. *Le Quotidien de l'Art*. Available at: http://www.lequotidiendelart.com/articles/8794-exclusif-apres-l-affaire-cranach-l-affaire-gentileschi-a-la-national-gallery.html [accessed 17 February 2017].
4. Sfez, Z. (2016). 'Une affaire de présumés faux tableaux ébranle le milieu de l'art au niveau mondial'. *Le Journal de la culture*. Available at: https://www.franceculture.fr/emissions/le-journal-de-la-culture/une-affaire-de-presumes-faux-tableaux-ebranle-le-milieu-de-lart [accessed 17 February 2017].
5. Bindé, J. (2016). 'Un Cranach saisi à Aix, et désormais un Gentileschi décroché de la National Gallery'. *Arts & Scènes*. Available at: http://www.telerama.fr/scenes/de-faux-tableaux-et-un-corbeau-peut-etre-enfin-demasque,139767.php [accessed 17 February 2017].
6. Sherwin, A. (2016). 'Art world rocked by 'genius' £200m Old Masters forgery scandal'. *i News*. Available at: https://inews.co.uk/essentials/culture/arts/art-world-rocked-genius-200m-old-masters-forgery-scandal/ [accessed 17 February 2017].

7. Noce, V. (2016). 'The man at the centre of the Old Master fakes scandal'. *The Art Newspaper*. Available at: http://authenticationinart.org/pdf/artmarket/the-centre-man.pdf [accessed 17 February 2017].
8. McEwan, O. (2016). 'Lessons from the unfolding Old Masters forgery Scandal'. *Hyperallergic*. Available at: http://hyperallergic.com/333140/lessons-from-the-unfolding-old-masters-forgery-scandal/ [accessed 17 February 2017].
9. Siegal, N. (2016). 'A dubious Old Master unnerves the art world'. *The New York Times*. Available at: https://www.nytimes.com/2016/10/27/arts/design/a-dubious-old-master-unnerves-the-art-world.html [accessed 17 February 2017].
10. Maneker, M. (2016). 'The Frans Hals revelations cast doubt on both technical expertise and the connoisseurship'. *Art /Market Monitor*. Available at: http://www.artmarketmonitor.com/2016/10/10/the-frans-hals-revelations-cast-doubt-on-both-technical-expertise-and-connoisseurship/ [accessed 17 February 2017].
11. BBC News (2016). 'Sotheby's declares 'Frans Hals' work a forgery'. Available at: http://www.bbc.com/news/entertainment-arts-37574411 [accessed 18 February 2017].
12. Gopnik, B. (2016). 'The latest forgery scandal is a tempest in a... fruit bowl'. *Artnet News*. Available at: https://news.artnet.com/opinion/frans-hals-forgery-701081 [accessed 18 February 2017].
13. Cascone, S. (2016). 'The Frans Hals forgery scandal, explained'. *Artnet News*. Available at: https://news.artnet.com/art-world/old-masters-forgery-scandal-facts-702065 [accessed 17 February 2017].
14. Chester, L. (2017). 'Sotheby's and dealer Mark Weiss head to court in Old Master dispute over Frans Hals portrait'. *Antiques Trade Gazette*. Available at: https://www.antiquestradegazette.com/news/2017/sothebys-and-dealer-mark-weiss-head-to-court-in-old-master-dispute-over-frans-hals-portrait/ [accessed 18 February 2017].
15. Boucher, B. (2017). 'Sotheby's sues London dealer Mark Weiss over Frans Hals forgery. It's one canvas in a massive Old Master forgery scandal'. *Artnet News*. Available at: https://news.artnet.com/art-world/sothebys-sues-weiss-hals-849560 [accessed 18 February 2017].
16. Cascone, S. (2016). 'Suspected $255 Million Old Master forgery scandal continues to rock the art world'. *Artnet News*. Available at: https://news.artnet.com/art-world/more-details-old-mastery-forgery-ring-695777 [accessed 18 February 2017].
17. Sherwin, A. (2016). 'Art world rocked by 'genius' £200m Old Masters forgery scandal'. i News. Available at: https://inews.co.uk/essentials/

culture/arts/art-world-rocked-genius-200m-old-masters-forgery-scandal/ [accessed 17 February 2017].

18. Watson, L. (2017). 'Vivedly-colourful 'Renaissance' masterpiece ruled a fake after experts find green pigment not invented until 20th century'. *The Telegraph*. Available at: http://www.telegraph.co.uk/news/2017/01/19/vividly-colourful-renaissance-masterpiece-ruled-fake-experts/ [accessed 18 February 2017].
19. Art Law & More (2017). 'Sotheby's sue over fake Parmigianino'. Available at: https://artlawandmore.com/2017/01/19/sothebys-sue-over-fake-parmigianino/ [accessed 18 February 2017].
20. Art History News (2016). 'Who was André Borie?' (ctd.). Available at: http://www.arthistorynews.com/articles/3908_Who_was_Andr%C3%A9_Borie? [accessed 18 February 2017].
21. Art Law Blog (2016). 'Fake Old Master painting uncovered in Europe raises fears of more sophisticated forgeries on the market'. Available at: http://grossmanllp.com/art-law-blog/2016/11/fake-old-master-painting-uncovered-in-europe-raises-fears-of-more-sophisticated-forgeries-on-the-market/ [accessed 18 February 2017].
22. O'Neil, S. and Jenkins, R. (2007). 'The £10m art collection that was forged by a family in their garden shed in Bolton'. *The Times*. Available at: http://www.thetimes.co.uk/article/the-pound10m-art-collection-that-was-forged-by-a-family-in-their-garden-shed-in-bolton-vshhxzq25rv [accessed 18 February 2017].

Jackson Pollock

1. Blouinartinfo International (2007) 'Disputed Pollocks sold to dealer Ronald Feldman'. Available at: http://www.blouinartinfo.com/news/story/267593/disputed-pollocks-sold-to-dealer-ronald-feldman [accessed 10 January 2017].
2. Mullins, J. (2016). 'Jackson Pollock and the Art of the Psyche'. Available at: http://elkodaily.com/lifestyles/chapter-jackson-pollock-and-the-art-of-the-psyche/article_10067b2d-98b6-5d94-b075-6aa55cfc5bbc.html [accessed 30 February 2017].
3. MacAdam, B. A. (2007). 'Top ten artnews stories: "Not a picture but an event"'. *Artnews*. Available at: http://www.artnews.com/2007/11/01/top-ten-artnews-stories-not-a-picture-but-an-event/ [accessed 25 February 2017].
4. Manoogian, D. (2007). 'Technical Analysis of Three Paintings Attributed to Jackson Pollock', *Harvard University Art Museums*, Technical Report.

5. Thermo Nicolet (2014). 'Introduction to Fourier Transform Infrared spectrometry'. Available at: https://www.google.com.eg/url?sa=t&rct=j&q=&esrc=s&source=web&cd=1&cad=rja&uact=8&ved=0ahUKEwiYhfOgh4fXAhXSZVAKHbaNACkQFggsMAA&url=https%3A%2F%2Fwww.researchgate.net%2Ffile.PostFileLoader.html%3Fid%3D57ce879ded99e1e0364482dd%26assetKey%3DAS%253A403246750420994%25401473152925890&usg=AOvVaw37zb_muZurKPXoOTDEaYkD [accessed 20 September 2014].
6. Kennedy, R. (2005). 'Is this a real Jackson Pollock?' *The New York Times*. Available at: http://www.nytimes.com/2005/05/29/arts/design/29kenn.html? pagewanted=all&_r=0 [accessed 13 July 2014].
7. The Arts Fuse (2008). 'Fuse flash: Anybody see the fat lady? Pollock Matter affair still gropes for that final act'. Available at: http://artsfuse.org/486/fuse-flash-anybody-see-the-fat-lady-pollock-matter-affair-still-gropes-for-that-final-act/ [accessed 14 July 2014].
8. The Art Law Blog (2007). 'Is Alex Matter prepared to concede the paintings are not authentic? No. We don't regard Mr. Martin's conclusions as reliable'. Available at: http://theartlawblog.blogspot.com/2007/12/is-alex-matter-prepared-to-concede.html [accessed 11 November 2014].
9. Nielson, M. R. (2011). 'Authentic or not? Chemistry solves the mystery'. Available at: https://www.acs.org/content/dam/acsorg/education/resources/highschool/chemmatters/videos/chemmatters-april2011-authentic-or-not.pdf [accessed 29 November 2014].
10. Norman, M. (2014). 'The story behind Pigment 254, nicknamed 'Ferrari Red''. Available at: http://blog.cleveland.com/pdextra/2007/10/pollock_cuts.html [accessed 13 July 2014].

AUTHENTICATION

Lucian Freud

1. Feaver, A. (2011). 'Lucian Freud: Friend, good cook and man of very rude letters'. *The Guardian*. Available at: https://www.theguardian.com/artanddesign/2011/jul/24/lucian-freud-william-feaver-appreciation [accessed 27 September 2016].
2. Cunningham, I. (2011). 'Lucian Freud dies aged 88'. *Harper's Bazaar UK*. Available at: http://www.harpersbazaar.co.uk/fashion/fashion-news/news/a5451/lucian-freud-dies-aged-88/ [accessed 27 September 2016].
3. Kamp, D. (2014). 'Freud, interrupted'. Available at: http://losarciniegas.blogspot.com.eg/2014/01/david-kamp-freud-interrupted.html [accessed 5 October 2016].

4. Filler, M. (2012). 'The naked and the ID'. *Vanity Fair*. Available at: http://www.vanityfair.com/culture/1993/11/freud-199311 [accessed 5 October 2016].
5. Moore, C. (2013). 'Portrait of Lucian Freud, a dangerous, devoted artist'. *The Telegraph*. Available at: http://www.telegraph.co.uk/culture/books/non_fictionreviews/10359496/Portrait-of-Lucian-Freud-a-dangerous-devoted-artist.html [accessed 27 September 2016].
6. Grimes, W. (2011). 'Lucian Freud, figurative painter who redefined Portraiture, is dead at 88'. *The New York Times*. Available at: http://www.nytimes.com/2011/07/22/arts/lucian-freud-adept-portraiture-artist-dies-at-88.html [accessed 5 October 2016].
7. Foster, P. (2016). 'Portrait that Lucian Freud denied painting revealed as genuine'. *The Telegraph*. Available at: http://www.telegraph.co.uk/news/2016/07/17/portrait-that-lucian-freud-denied-painting-revealed-as-genuine/ [accessed 28 September 2016].
8. Daily Mail Reporter (2016). 'Did Lucian Freud lie for decades that this painting wasn't his... so rival he loathed couldn't sell it?'. *Mail online*. Available at: http://www.dailymail.co.uk/news/article-3685564/Did-Lucian-Freud-lie-decades-painting-wasn-t-rival-loathed-couldn-t-sell-it.html [accessed 28 September 2016].
9. Hodges, M. (2016). 'Fake or Fortune? Answers a £500,000 question: Is this a genuine Lucian Freud?' *Radiotimes*. Available at: http://www.radiotimes.com/news/2016-07-17/fake-or-fortune-answers-a-500000-question-is-this-a-genuine-lucian-freud [accessed 28 September 2016].
10. BBC (2016). 'Fake or Fortune? Series 5, Freud'. *Youtube*. Available at: https://www.youtube.com/watch?v=Dk6gXlmlFfo [accessed 30 September 2016].
11. Sheldon, L. University College London, email sent to Jehane Ragai, 20 December 2016.
12. Grosvenor, B. (2016). 'Early Freud revealed on 'Fake of Fortune?'' *Art History News*. Available at: http://www.arthistorynews.com/articles/4041_Early_Freud_revealed_on_Fake_of_Fortune [accessed 30 September 2016].

Edouard Manet

1. McCrone, W. C. (2001). 'Artful dodgers: Virtuosos of art forgery meet the masters of scientific detection', *The Sciences*, **41**(1), pp. 32–37.

2. Brainerd, A. W. (1988). *The Infanta Adventure and the Lost Manet*, Reichl Press, Indiana.
3. Boime, A. and Kossolapov A. (2003). 'Manet's lost infanta', *Journal of the American Institute for Conservation*, **42**(3), pp. 407–418.

Mahmoud Said

1. Rashed, D. (2007). 'The real McCoy'. *AlAhram Weekly On-line, Issue 869*. Available at: http://weekly.ahram.org.eg/archive/2007/869/fe1.htm [accessed 30 August 2014].
2. Kanafani, F. M. (2015). 'Buyer beware, part 2: The way forward'. *Mada Masr*. Available at: www.madamasr.com/en/2015/05/27/feature/.../buyer-beware-part-2-the-way-forward/ [accessed 14 December 2016].
3. Lamiz. 'Mahmoud Said — An artist that revolutionized modern Egyptian art'. Available at: http://www.lamiz.com/travel/mahmoud-said-artist-revolutionized-modern-egyptian-art.html [accessed 14 December 2016].
4. Wikipedia (2016). 'Mahmoud Sa'id'. Available at: https://en.wikipedia.org/wiki/Mahmoud_Sa'id [accessed 14 December 2016].
5. Al-Ahram Weekly (2014). 'The new Mahmoud Said Museum'. Available at: http://weekly.ahram.org.eg/News/7771/-/Letter.aspx [accessed 14 December 2016].
6. Wife of a former Egyptian Ambassador to the United Nations, personal communication, November 2012.

Vincent van Gogh

1. Van Gogh Museum (2013). 'The letters'. Available at: http://vangoghletters.org/vg/letters.html [accessed 9 November 2013].
2. Dik, J., Janssens, K., Van DerSnickt, G., Van DerLoeff, L., Rickers, K. and Cotte, M. (2008). 'Visualization of a lost painting by Vincent Van Gogh using synchrotron based X-ray fluorescence elemental mapping', *Analytical Chemistry*, **80**(16), pp. 6436–6442.
3. Van Tilborgh, L., Meedendorp, T. and van Maanen, O. (2013). 'Sunset at Montmajour: A newly discovered painting by Vincent van Gogh', *Burlington Magazine*, **155**(1327), pp. 696–705.
4. Siegal, N. (2013). 'A van Gogh's trip from the attic to the museum'. *New York Times*. http://www.nytimes.com/2013/09/10/arts/design/new-van-gogh-painting-discovered-in-amsterdam.html [accessed 9 November 2013].

5. Clark, N. (2013). 'Sunset at Montmajour: 'Sensational' Van Gogh work discovered 20 years after experts rejected it as a fake'. *The Independent*. Available at: http://www.independent.co.uk/arts-entertainment/art/news/sunset-at-montmajour-sensational-van-gogh-work-discovered-20-years-after-experts-rejected-it-as-a-8804681.html [accessed 9 November 2013].
6. Brown, M. (2013). 'Newly discovered Van Gogh painting kept in Norwegian attic for years'. *The Guardian*. Available at: http://www.theguardian.com/artanddesign/2013/sep/09/van-gogh-painting-discovered [accessed 9 November 2013].
7. Bailey, M. (2013). 'Van Gogh's Sunset at Montmajour: 'A major addition to the oeuvre'. *The Telegraph*. Available at: http://www.telegraph.co.uk/culture/art/10299372/Van-Goghs-Sunset-at-Montmajour-a-major-addition-to-the-oeuvre.html [accessed 18 November 2013].
8. Hughes, O. (2013). 'Lost Vincent Van Gogh painting 'Sunset at Montmajour' found in a Norwegian attic'. Available at: http://nerdalicious.com.au/art/lost-vincent-van-gogh-painting-sunset-at-montmajour-found-in-a-norwegian-attic/ [accessed 9 November 2013].

Rembrandt van Rijn

1. Cruwys, B. (2013). 'Lost painting by the Dutch master Rembrandt discovered in Devon.' Available at: http://www.itv.com/news/westcountry/2013-03-18/lost-painting-by-the-dutch-master-rembrandt-discovered-in-devon/ [accessed 12 August 2014].
2. BBC News Devon Arts (2014). 'Buckland Abbey Rembrandt self-portrait is genuine'. Available at: http://www.bbc.com/news/uk-england-devon-27763664 [accessed 12 August 2014].
3. Sky News (2014). 'Rembrandt self-portrait worth £30m is verified'. Available at: http://news.sky.com/story/rembrandt-self-portrait-worth-30m-is-verified-10401536 [accessed 11August 2014].
4. ITV (2014). 'Self-portrait verified as a Rembrandt after months of scientific analysis'. Available at: http://www.itv.com/news/2014-06-10/self-portrait-verified-as-a-rembrandt-after-months-of-scientific-analysis/ [accessed 11 August 2014].
5. Cambridge News (2014). 'Self-portrait verified as Rembrandt by Cambridge University's Whittlesford-based Hamilton Kerr Institute'. Available at: http://www.cambridge-news.co.uk/Selfportrait-verified-Rembrandt-Cambridge-Universitys-Whittlesfordbased-Hamilton-Kerr-Institute/story-22369884-detail/story.html [accessed 11 August 2014].
6. Furness, H. (2014). 'Self portrait is authentic Rembrandt, National Trust confirm'. *The Telegraph*. Available at: http://www.telegraph.co.uk/culture/art/art-news/10886646/Self-portrait-is-authentic-Rembrandt-National-Trust-confirm.html [accessed 11 August 2014].

Johannes Vermeer

1. Christie's catalogue (2014). 'Johannes Vermeer Saint Praxedis'. Available at: http://www.christies.com/presscenter/pdf/2014/CATALOUGE_NOTE_Johannes_Vermeer_Delft_1632_1675_Saint_Praxedis_lot_39.pdf [accessed 30 September 2014].
2. The History Blog (2014). 'One of Vermeer's first paintings authenticated Saint Praxedis'. Available at: http://www.thehistoryblog.com/archives/30929 [accessed 30 September 2014].
3. Wheelock, A. K. (1986). 'St. Praxedis: New light on the early career of Vermeer'. *Artibuset Historiae,* **17**(14), pp. 71–89.
4. David, N. G. (2014). 'Vermeer painting 'Saint Praxedis' sells for more than $10 million'. *Los Angeles Times*. Available at: http://www.latimes.com/entertainment/arts/culture/la-et-cm-vermeer-painting-auction-20140708-story.html [accessed 29 September 2014].

Leonardo da Vinci

1. Jones, J. (2011). 'Leonardo's 'lost' Christ, sold for £45 in 1956'. Available at: http://www.theguardian.com/artanddesign/2011/jul/12/leonardo-portrait-of-christ-salvator-mundi [accessed 5 December 2014].
2. Antiques Trade Gazette (2011). 'National Gallery approves new Leonardo discovery'. Available at: https://www.antiquestradegazette.com/news/2011/national-gallery-approves-new-leonardo-discovery/ [accessed 5 December 2014].
3. Esterow, M. (2011). 'Updated: A long lost Leonardo'. Available at: http://www.artnews.com/2011/08/15/updated-a-long-lost-leonardo-2/ [accessed 15 October 2014].
4. Kemp, M. (2011). 'Art history: Sight and salvation', *Nature,* **479**(7372), pp. 174–175.
5. Artdaily (2014). 'Da Vinci discovered: Painting gains attribution after careful scholarship and conservation'. Available at: http://artdaily.com/news/48943/Da-Vinci-Discovered--Painting-Gains-Attribution-After-Careful-Scholarship-and-Conservation#.VJk5gf90Bg [accessed 15 October 2014].
6. Allsop, L. (2011). 'Decoding da Vinci: How a lost Leonardo was found'. Available at: http://edition.cnn.com/2011/11/04/living/discovering-leonardo-salvator-mundi/ [accessed 10 November 2014].
7. Hudson, M. (2017). 'There are 450 million reasons why Leonardo da Vinci's *Salvator Mundi* isn't a masterpiece'. Available at: http://www.telegraph.co.uk/art/artists/450-million-reasons-leonardo-da-vincis-salvator-mundi-isnt-masterpiece/ [accessed 15 November 2017].

PART II

THE COURT

Elena Basner: The Expert Under Siege

1. RusArtNet (2015). 'Worldbackwards: The Elena Basner case'. Available at: http://www.rusartnet.com/russia/history/modern/scandals/art-forgery/worldbackwards-the-elena-basner-case [accessed 18 May 2016].
2. Lindsay, I. (2014). 'Yelena Basner's house arrest is extended'. Available at: http://www.russianartdealer.com/journal/yelena-basners-house-arrest-extended/#sthash.jkV3s8I6.dpu [accessed 18 May 2016].
3. Lindsay, I. (2014). 'Avant-garde expert Yelena Basner arrested'. Available at: http://www.russianartdealer.com/journal/avant-garde-expert-yelena-basner-arrested/ [accessed 18 May 2016].
4. Art Media Agency (2016). 'Russian art historian on trial over forgeries'. Available at: http://en.artmediaagency.com/tag/forgery/ [accessed 18 May 2016].
5. Russian News Online (2016). 'The court acquitted the critic Elena Basner the case of Fraud-RBC'. Available at: http://russiannewsonline.blogspot.com.eg/2016/05/the-court-acquitted-critic-elena-basner.html [accessed 18 May 2016].
6. Trifonov, V. and Pushkarskaya, A. (2012). 'Art Collector Accuses Russian Museum of Forgery'. Available at: http://www.worldcrunch.com/culture-society/art-collector-accuses-russian-museum-of-forgery/russia-painting-forgery-boris-grigoriev-soviet-union-saint-petersburg-museum/c3s9353/ [accessed 18 May 2016].
7. Avrora Fine Arts (2012). 'Investment Ltd *vs* Christie, Manson & Woods Ltd'. Available at: http://authenticationinart.org/pdf/artlaw/20120727.pdf [accessed 5 November 2013].
8. Akinsha, K., Hochfield, S., Artemiev, Z., Fitzgerald, N. and Rump, G. C. (2009). 'The Faking of Russian Avant-Garde'. Available at: http://www.artnews.com/2009/07/01/the-faking-of-the-russian-avant-garde/ [accessed 11 August 2014].
9. Telekhov, M. (2016). 'Prominent Russian art expert acquitted of fraud'. Available at: http://www.rapsinews.com/judicial_news/20160517/276118059.html [accessed 20 May 2016].
10. The Newspapers (2016). 'The court acquitted the critic Elena Basner in the case of fraud'. Available at: http://the-newspapers.com/2016/05/17/the-court-acquitted-the-critic-elena-basner-in-the-case-of-fraud [accessed 20 May 2016].

11. The Newspapers (2016). 'The victim of the critic Basner acquittal: falsifiers covered felon'. Available at: http://the-newspapers.com/2016/05/17/the-victim-of-the-critic-basner-acquittal-falsifiers-covered-felon [accessed 20 May 2016].
12. Helmer, J. (2015). 'The great falshak'. Available at: http://www.counterpunch.org/2015/05/01/the-great-falshak/ [accessed 20 May 2016].

Plight of the Auction House

1. Heddaya, M. (2015). 'Will the Sotheby's Caravaggio Decision Impact the Practice of Authentication?' *Blouinartinfo*. Available at: http://www.blouinartinfo.com/news/story/1073843/will-the-sothebys-caravaggio-decision-impact-the-practice-of [accessed 1 June 2016].
2. Goldstein, J. L. (2011). 'The card players of Caravaggio, Cézanne and Mark Twain: Tips for getting lucky in high-stakes research', *Nature Medicine*, **17**(10), pp. 1201–1205.
3. Croft, J. (2015). 'Judge backs Sotheby's in Caravaggio legal battle'. *Financial Times*. Available at: http://www.ft.com/cms/s/0/685030ba-9d6f-11e4-8946-00144feabdc0.html#axzz4JlAd26Cj [accessed 1 June 2016].
4. Hawkins, R. (2015). 'Thwaytes v Sotheby's: Caravaggio's Cardsharps?' *Privateartinvestor*. Available at: http://www.privateartinvestor.com/art-law-2/thwaytes-v-sothebys-caravaggio-cardsharps/ [accessed 27 February 2016].
5. Capon, A. (2014). 'Caravaggio case — Sotheby's come out fighting'. *Antiques Trade Gazette*. Available at: https://www.antiquestradegazette.com/news/2014/caravaggio-case-sotheby-s-come-out-fighting/ [accessed 30 March 2016].
6. Mirror (2014). 'Sotheby's, sued after £10m Caravaggio original sold for just £42k'. Available at: http://www.mirror.co.uk/news/uk-news/sothebys-sued-after-10m-caravaggio-4519421 [accessed 19 February 2016].
7. Shirbon, E. (2015). 'Sotheby's fends off rare negligence suit over Caravaggio dispute'. *Reuters*. Available at: http://www.reuters.com/article/us-britain-caravaggio-idUSKBN0KP1LQ20150116 [accessed 27 March 2016].
8. England and Wales High Court (2015). 'Between: Mr. Lancelot Thwaytes and Sotheby's'. Available at: http://authenticationinart.org/pdf/art-law/Thwaytes-v-Sothebys1.pdf [accessed 23 May 216].
9. Art History News (2015). 'Caravaggio's lost 'Card Sharps'? (ctd.). Available at: http://www.arthistorynews.com/articles/3202_Caravaggios_lost_Card_Sharps_ctd [accessed 18 February 2016].

10. Spencer, R. D. (2015). 'What is the art law on authenticating works by old masters?' *Artnet News*. Available at: https://news.artnet.com/art-world/how-do-experts-arrive-at-an-opinion-about-artwork-authenticity-323652 [accessed 20 February 2016].
11. Furness, H. (2015). 'Caravaggio in court: Sotheby's wins case after '£10m' painting sold for £42,000'. *The Telegraph*. Available at: http://www.telegraph.co.uk/news/uknews/law-and-order/11349854/Caravaggio-in-court-Sothebys-wins-case-after-10m-painting-sold-for-42000.html [accessed 22 May 2016].
12. CBM & Partners (2015). 'SOTHEBY'S V. THWAYTES -Authenticity issues and intermediary's liability-'. Available at: http://www.cbmlaw.it/sothebys-v-thwaytes-authenticity-issues-intermediarys-liability/ [accessed 21 May 2016].
13. Perez, M. (2014). 'The 5 most controversial legal battles over art'. *The Richest*. Available at: http://www.therichest.com/rich-list/most-influential/the-5-most-controversial-legal-battles-over-art/ [accessed 25 May 2016].
14. DuBoff, L. D. and King, C. O. (2006). *Art Law in a Nutshell*, West Group, USA, pp. 61–67.

Plight of the Collector

1. Kinsella, E. (2016). 'Sparks fly at Knoedler trial as defrauded buyer of fake Rothko painting takes stand'. *Artnet News*. Available at: https://news.artnet.com/market/knoedler-trial-day-3-defrauded-rothko-buyer-takes-stand-415271 [accessed 4 June 2016].
2. Ward, V. (2016). 'Inside the trial all the billionaires are watching'. *Town & Country*. Available at: http://www.townandcountrymag.com/society/money-and-power/news/a4965/knoedler-trial/ [accessed 4 June 2016].
3. Miller, M. H. (2016). 'Tears, shouting matches, and a slew of expert witnesses as Knoedler trial heads into week two'. *Artnews*. Available at: http://www.artnews.com/2016/02/01/tears-shouting-matches-and-a-slew-of-expert-witnesses-as-knoedler-trial-heads-into-week-two/ [accessed 3 June 2016].
4. Kahn , D. (2016). 'More than money: Inside the Knoedler art fakes scandal'. *Observer*. Available at: http://observer.com/2016/02/more-than-money-inside-the-knoedler-art-fakes-scandal/ [accessed 15 May 2016].
5. Munro, C. (2016). 'Conservator found Rothko painting in Knoedler trial to be a 'deliberate fake''. *Artnet News*. Available at: https://news.artnet.com/art-world/former-librarian-hot-seat-knoedler-fraud-trial-420161 [accessed 20 May 2016].

6. Heddaya, M. (2013). 'Knoedler gallery canoodler Glafira Rosales arrested for allegedly hiding $12.5M'. *Hyperallergic*. Available at: http://hyperallergic.com/71658/knoedler-Gallery-canoodler-glafira-rosales-arrested-for-hiding-12-5m/ [accessed 15 May 2016].
7. Charney, N. (2015). *The Art of Forgery: The Minds, Motives and Methods of the Master Forgers*, Phaidon Press, pp. 159–164.
8. Flescher, S. (2016). 'Letter to the editor: From Sharon Flescher, IFAR'. *The Art Newspaper*. Available at: http://old.theartnewspaper.com/comment/letter-to-the-editor-from-sharon-flescher-ifar/ [accessed 6 June 2016].
9. Shnayerson, M. (2012). 'A question of provenance'. *Vanity Fair*. Available at: http://www.vanityfair.com/culture/2012/05/knoedler-Gallery-forgery-scandal-investigation [accessed 6 June 2016].
10. Lescaze, Z. (2013). 'Glafira Rosales, dealer tied to allegedly fake Knoedler paintings, charged with tax fraud, concealing $12.5 M in proceeds'. *Observer*. Available at: http://observer.com/2013/05/dealer-glafira-rosales-charged-with-tax-fraud-concealing-12-5-m-in-proceeds/ [accessed 15 May 2016].
11. Corbett, R. (2012). 'Knoedler Gallery settles lawsuit with Pierre Lagrange over 'fake' Pollock'. *Blouinartinfo*. Available at: http://blogs.artinfo.com/artintheair/2012/10/05/knoedler-Gallery-settles-lawsuit-with-pierre-lagrange-over-fake-pollock/ [accessed 15 May 2016].
12. Munro, C. (2016). 'The 5 best moments of the Knoedler fraud trial so far'. *Artnet News*. Available at: https://news.artnet.com/art-world/top-5-knoedler-trial-421031 [accessed 4 June 2016].
13. Miller, M. H. (2016). 'It's over: Knoedler settles with the De Soles, concluding three-week case over $8.3 m. fake Rothko'. *Artnews*. Available at: http://www.artnews.com/2016/02/10/its-over-knoedler-settles-with-the-de-soles-concluding-three-week-case-over-8-3-m-fake-rothko/ [accessed 4 June 2016].
14. Nir, S. M., Cohen, P. and Rashbaum, W. K. (2013). 'Struggling immigrant artist tied to $80 million New York fraud'. *The New York Times*. Available at: http://www.nytimes.com/2013/08/17/nyregion/struggling-immigrant-artist-tied-to-80-million-new-york-fraud.html [accessed 6 June 2016].
15. Miller, M. H. (2016). 'Ann Freedman settles with the De Soles in Knoedler trial over $8.3 M. fake Rothko'. *Artnews*. Available at: http://www.artnews.com/2016/02/07/ann-freedman-settles-with-the-de-soles-in-knoedler-trial/ [accessed 15 May 2016].
16. Kinsells, E. (2016). 'Richard Diebenkorn's Daughter Challenges Ann Freedman's Story at Knoedler Forgery Trial'. *Artnet News*. Available at: https://news.artnet.com/market/gretchen-diebenkorn-knoedler-forgery-trial-414617 [accessed 6 June 2016].

17. Miller, M. H. (2016). 'The big fake: behind the scenes of Knoedler Gallery's downfall'. *Artnews*. Available at: http://www.artnews.com/2016/04/25/the-big-fake-behind-the-scenes-of-knoedler-Gallerys-downfall/ [accessed 6 June 2016].
18. Gilbert, L. (2016). 'Red flags were flying' around Knoedler fakes, experts testify'. *AiA Art News-service*. Available at: http://authenticationinart.org/pdf/artmarket/red-flags-knoedler.pdf [accessed 24 May 2016].
19. Kinsella, E. and Cascone, S. (2016). 'Top 9 takeaways from Knoedler forgery trial'. *Artnet News*. Available at: https://news.artnet.com/exhibitions/top-takeaways-from-knoedler-forgery-trial-426086 [accessed 4 June 2016].
20. Rashbaum, W. K. and Cohen, P. (2013). 'Art dealer admits to role in fraud'. *The New York Times*. Available at: http://www.nytimes.com/2013/09/17/arts/design/art-dealer-admits-role-in-selling-fake-works.html [accessed 6 June 2016].
21. Plous, S. (1993). *The Psychology of Judgment and Decision Making*, McGraw-Hill, USA, p. 233.
22. Wikipedia (2016). 'Conformation bias'. Available at: https://en.wikipedia.org/wiki/Confirmation_bias [accessed 6 June 2016].
23. Devey, J. (ed.) (1902). Sir Francis Bacon. *Novum Organon*, P. F. Collier, New York, p. 24.
24. Gilbert, L. and Glass, B. (2016). 'Former director of scandal-beset Knoedler Gallery breaks her silence'. *The Art Newspaper*. Available at: http://old.theartnewspaper.com/news/news/former-director-of-scandal-beset-knoedler-gallery-breaks-her-silence/ [accessed 6 June 2016].
25. Gilbert, L. (2015). 'Five legal cases changing the art market as we know it'. *Artsy*. Available at: https://www.artsy.net/article/artsy-editorial-five-legal-cases-changing-the-art-market-as-we-know-it [accessed 24 May 2016].

SAFEGUARDING THE ART MARKET

1. Interview with Robert Norton (2016). 'Blockchain technology and the application to the art market'. *Art and Finance Report*. Available at: https://www2.deloitte.com/content/dam/Deloitte/lu/Documents/financial-services/artandfinance/lu-en-artandfinancereport-21042016.pdf [accessed 10 January 2017].
2. Ray, K. P. (2015). 'New York senate passes bill to protect art authenticators'. *Cultural Assets*. Available at: http://www.gtlaw-culturalassets.com/2015/06/new-york-senate-passes-bill-to-protect-art-authenticators/ [accessed 2 March 2016].

3. Jasani, A. (2015). 'A New York state bill seeking to protect art authenticators'. *Art@Law*. Available at: http://www.artatlaw.com/archives/a-new-york-state-bill-seeking-to-protect-art-authenticators [accessed 2 March 2016].
4. Tarsis, I. (2015). 'Hold your horses: Art authenticators not protected yet'. *Center for Art Law*. Available at: https://itsartlaw.com/2015/07/01/hold-your-horses-authenticators/ [accessed 2 March 2016].
5. Authentication in Art (2016). 'AiA Art and Law'. Available at: http://authenticationinart.org/aia-archive/art-law/ [accessed 23 June 2016].
6. Karczewski, L. (2015). 'Using bioengineered DNA as a tool for authenticating art'. *Art Law*. Available at: https://artlaw.foxrothschild.com/2015/10/articles/art-authentication/using-bioengineered-dna-as-a-tool-for-authenticating-art/ [accessed 2 July 2016].
7. Woollaston, V. (2016). 'Good luck faking that! Artwork will soon be 'tagged' with DNA to help experts identify forgeries'. *Mail Online*. Available at: http://www.dailymail.co.uk/sciencetech/article-3269136/Good-luck-faking-Artwork-soon-tagged-artist-s-DNA-help-experts-identify-forgeries.html [accessed 2 July 2016].
8. i2M standards (2015). Available at: https://www.i2mstandards.org/ [accessed 14 July 2016].
9. Campbell-Dollaghan, K. (2015). 'Forgery is getting so good that scientists had to invent keys made from DNA'. *GIZMODO*. Available at: http://gizmodo.com/forgery-is-getting-so-good-that-scientists-had-to-inven-1736335563 [accessed 10 July 2016].
10. Karczewski, L. (2015). 'Authenticating art with bioengineered DNA: The IP issues'. *Law 360*. Available at: http://www.aristitle.com/news/docs/Authenticating%20Art%20With%20Bioengineered%20DNA%20The%20IP%20Issues_110215.pdf [accessed 17 July 2016].
11. Tapscott, D. and Tapscott, A. (2016). *Blockchain Revolution: How the Technology Behind Bitcoin is Changing Money, Business and the World*. Penguin Random House.
12. Custodio, N. (2014). 'Still don't get Bitcoin? Here's an explanation even a five-year-old will understand'. *CoinDesk*. Available at: http://www.coindesk.com/bitcoin-explained-five-year-old/ [accessed 10 January 2017].
13. Perez, Y. B. (2015). 'How Blockchain Tech is inspiring the art world'. *CoinDesk*. Available at: http://www.coindesk.com/blockchain-technology-inspiring-art/ [accessed 10 January 2017].
14. Michalska, J. (2016). 'Blockchain: How the revolutionary technology behind Bitcoin could change the art market'. *The Art Newspaper*. Available at: http://old.theartnewspaper.com/news/news/blockchain-how-the-revolutionary-technology-could-change-the-art-world/ [accessed 11 January 2017].

15. Johnson, P. and Mcnamara, R. (2015). 'The Blockchain and digital art: On ascribe, and its new form of art authentication'. *Art F City*. Available at: http://artfcity.com/2015/10/22/the-blockchain-and-digital-art-on-ascribe-and-its-new-form-of-art-authentication/ [accessed 11 January 2017].
16. Butcher, M. (2015). 'Verisart plans to use the blockchain to verify the authenticity of artworks'. *Techcrunch*. Available at: https://techcrunch.com/2015/07/07/verisart-plans-to-use-the-blockchain-to-verify-the-authencity-of-artworks/ [accessed 11 January 2017].

PART III

LA BELLA PRINCIPESSA

A Possible Solution to the Enigma

1. William Hazlitt quote. Available at http://quotes.yourdictionary.com/author/william-hazlitt/61013 [accessed 2 May 2017].
2. Dorment, R. (2010). 'La Bella Principessa: A $100m Leonardo, or a copy?' *The Telegraph*. Available at: http://www.telegraph.co.uk/search/?queryText=La+Bella+Principessa%3A+a+%24100m+Leonardo%2C+or+a+copy%3F&sort=recent [accessed 1 September 2014].
3. Kemp, M. and Cotte, P. (2010). *'La Bella Principessa: The Story of the New Masterpiece by Leonardo da Vinci'*, Hodder& Stoughton Ltd., London, p. 98.
4. Esterow, M. (2010). 'The real thing?' *Artnews*. Available at: http://www.artnews.com/2010/01/01/the-real-thing/ [accessed 6 August 2016].
5. O'Neill, T. (2012). 'Lady with a secret'. *National Geographic*. Available at: http://ngm.nationalgeographic.com/2012/02/lost-da-vinci/o-neill-text [accessed 10 March 2015].
6. Cotte, P. and Kemp, M. (2014). 'La Bella Principessa and the Warsaw Sforziada'. *Lumiere Technology*. Available at: http://www.lumiere-technology.com/news/Study_Bella_Principessa_and_Warsaw_Sforziad.pdf [accessed 14 September 2014].
7. Wax, R. (2013). '*De Divina Proportione* is a book by Luca Pacioli illustrated by Leonardo da Vinci'. Available at: http://www.graphicine.com/a-picture-is-worth/ [accessed 14 September 2014].
8. Welsh, J. (2011). 'Lost' Da Vinci portrait, and its origins, stir controversy'. *LiveScience*. Available at: http://www.livescience.com/16549-lost-davinci-portrait.html [accessed 14 September 2014].
9. Noce, V. (2016). 'La Bella Principessa: Still an Enigma'. *The Art Newspaper*. Available at: http://old.theartnewspaper.com/features/la-bella-principessa-still-an-enigma/ [accessed 10 June 2016].

10. Jones, J. (2015). 'This is a Leonardo da Vinci? The gullible experts have been duped again. *The Guardian*. Available at: https://www.theguardian.com/commentisfree/2015/nov/30/leonardo-da-vinci-experts-painting-la-bella-principessa [accessed 12 December 2015].
11. Daley, M. (2012). 'Problems with 'La Bella Principessa' — Part I: The Look'. *ArtWatch UK Online*. Available at: http://artwatch.org.uk/problems-with-la-bella-principessa-part-i-the-look-2/ [accessed 10 June 2016].
12. Daley, M. (2016a). 'Problems with 'La Bella Principessa'' — Part II: Authentication Crisis'. *ArtWatch UK Online*. Available at: http://artwatch.org.uk/problems-with-la-bella-principessa-part-ii-authentication-crisis/ [accessed 10 June 2016].
13. Krzyżagórska-Pisarek, K. (2015). 'La Bella Principessa. Arguments against the Attribution to Leonardo'. *ArtWatch UK Online*. Available at: http://artwatch.org.uk/wp-content/uploads/2016/04/Leonardo-Krzyzagorska-Katarzyna_07-05-2015.pdf [accessed 10 June 2016].
14. Daley, M. (2016b). 'Problems with 'La Bella Principessa' — Part III: Dr. Pisarek responds to Prof. Kemp'. *ArtWatch UK Online*. Available at: http://artwatch.org.uk/problems-with-la-bella-principessa-part-iii-dr-pisarek-responds-to-prof-kemp/ [accessed 10 June 2016].
15. Kemp, M. (2015). 'Leonardo da Vinci La Bella Principessa. Errors, Misconceptions and Allegations of Forgery'. Available at: http://authenticationinart.org/pdf/literature/la-bella-principessa.pdf [accessed 10 June 2016].
16. Kemp, M. (2015). 'Reply (I) on 'La Bella Principessa' by Leonardo da Vinci'. Available at: http://www.academia.edu/19481859/Martin_Kemp_Reply_I_on_La_Bella_Principessa_by_Leonardo_da_Vinci_december_2015_ [accessed 10 June 2016].
17. Muñoz-Alonso, L. (2015). 'Forger Claims Leonardo da Vinci's *La Bella Principessa* is Actually His Painting of a Supermarket Cashier' *Artnet News*. Available at: https://news.artnet.com/art-world/forger-leornardo-da-vinci-bella-principessa-373451 [accessed 6 August 2016].
18. Newton, J. (2015). 'Is this Da Vinci masterpiece really just Sally the shop-girl from Bolton? Prolific forger claims he created the £100m artwork in based on a worker at his local Co-op'. *Mail Online*. Available at: http://www.dailymail.co.uk/news/article-3338200/Is-Da-Vinci-masterpiece-really-just-Sally-shopgirl-Bolton-Prolific-forger-claims-drew-100m-Renaissance-masterpiece-1970s-based-till-girl-local-op.html [accessed 6 August 2016].
19. Mount, H. (2015). '£100 million Leonardo — or checkout girl from Bolton? How British forger claims he fooled world with picture of

shopworker called Sally'. *Mail Online*. Available at: http://www.dailymail.co.uk/news/article-3338889/100-million-Leonardo-checkout-girl-Bolton-British-forger-claims-fooled-world-picture-shopworker-called-Sally.html [accessed 6 August 2016].

20. Austin, H. (2015). 'Scientists claim a forger's 'fake' Leonardo da Vinci painting is genuine'. *International Business Times*. Available at: http://www.ibtimes.co.uk/scientists-claim-forgers-fake-leonardo-da-vinci-painting-genuine-1531032 [accessed 7 August 2016].
21. Boswell, J. and Rayment, T. (2015). 'It's not a da Vinci, it's Sally from the Co-op'. *The Sunday Times*. Available at: http://www.thesundaytimes.co.uk/sto/news/uk_news/Arts/article1639169.ece [accessed 7 August 2016].
22. Art and Artifice (2015). 'La Bella Principessa: a Da Vinci or a copy?' Available at: http://aandalawblog.blogspot.com.eg/2015/12/la-bella-principessa-da-vinci-or-copy.html [accessed 8 August 2016].
23. Nicholas, S. (2015). 'Is this Leonardo da Vinci drawing a £100m masterpiece…or a worthless forgery'. Available at: http://www.express.co.uk/news/history/623077/Leonardo-da-Vinci-La-Bella-Principessa [accessed 8 June 2016].
24. Howard, K. (2015). 'Notorious art forger claims "La Bella Principessa" as his own'. *Melville House*. Available at: http://www.mhpbooks.com/notorious-art-forger-claims-la-bella-principessa-as-his-own/ [accessed 8 August 2016].
25. Universitá degli Studi di Pavia, Dipartimento di Chimica Generale (2011). *Misure del 210Pb mediante Spettrometria Gamma Diretta*, p. 48.
26. Kemp, M. (2016). 'Leonardo La Bella Principessa: personalization of the debate'. *Martin Kemp's This and That*. Available at: http://martinkempsthisandthat.blogspot.com.eg/2016/05/leonardo-la-bella-principessa.html [accessed 8 August 2016].
27. Gauthier, M. (1958). 'The psychology of the compulsive forger'. *Canadian Journal Corrections, Heinonline*. Available at: http://heinonline.org/HOL/LandingPage?handle=hein.journals/cjccj1&div=44&id=&page= [accessed 3 January 2014].
28. Ford, C. V. (1999). *Lies, Lies, Lies, The Psychology of Deceit*. American Psychiatric Publishing, p. 124.
29. Ragai, J. (2015). *The Scientist and the Forger*. Imperial College Press, London, p. 189.
30. Lopez J. (2008). *The Man Who Made Vermeers: Unvarnishing the Legend of Master Forger Han van Meegeren*. Houghton Mifflin Harcourt, Boston, p. 2.

31. Dutton, D. (ed.) (1983). *The Forger's Art: Forgery and the Philosophy of Art*, University of California Press, Berkeley, p. 5.
32. Forgy, M. (2012). *The Forger's Apprentice: Life with the World's Most Notorious Artist*. Create Space Independent Publishing Platform.
33. Irving, C. (1999). *Fake! The Story of Elmyr de Hory: The Greatest Art Forger of Our Time*. McGraw-Hill, New York.
34. Forgy, M. Writer and live-in assistant to Elmyr. E-mail sent to Jehane Ragai, 3 March 2015.

MONA LISAS

1. Wilde, O. 'Quotes: Quotable Quote'. *Goodreads*. Available at: http://www.goodreads.com/quotes/558084-imitation-is-the-sincerest-form-of-flattery-that-mediocrity-can [accessed 15 December 2016].
2. Gurstein, R. (2002). 'The mystic smile'. *New Republic*. Available at: https://newrepublic.com/article/122588/mystic-smile [accessed 11 October 2017].
3. Scailliérez, C. (2014). 'Mona Lisa – Portrait of Lisa Gherardini, wife of Francesco del Giocondo'. *Louvre*. Available at: http://www.louvre.fr/en/oeuvre-notices/mona-lisa-%E2%80%93-portrait-lisa-gherardini-wife-francesco-del-giocondo [accessed 11 June 2014].
4. Kemp, M. and Pallanti, G. (2017). *Mona Lisa: The People and the Painting*. Oxford University Press, p. 35.
5. Treasures of the world (2014). 'Theft of the Mona Lisa'. Available at: http://www.pbs.org/treasuresoftheworld/a_nav/mona_nav/main_monafrm.html [accessed 11 June 2014].
6. King, S. 'Apollinaire, Picasso, Mona Lisa'. *Today in Literature*. Available at: http://www.todayinliterature.com/stories.asp?Event_Date=9/7/1911 [accessed 5 January 2017].
7. deViguerie, L., Walter, P., Laval, E., Mottin, B. and Solé, V. A. (2010). 'Revealing the sfumato Technique of Leonardo da Vinci by X-Ray Fluorescence Spectroscopy', *Angewandte Chemie*, **122**(35), pp. 6261–6264.
8. Elias, M. and Cotte, P. (2008). 'Multispectral camera and radiative transfer equation used to depict Leonardo's sfumato in Mona Lisa', *Applied Optics*, **47**(12), pp. 2146–2154.
9. Dolan, A. (2010). 'French scientists crack secrets of 'Mona Lisa'. Available at: http://archive.boston.com/news/science/articles/2010/07/16/french_scientists_crack_secrets_of_mona_lisa/ [accessed 10 May 2014].
10. Ware, D. G. (2015). ''Mona Lisa' not 'Lisa' at all, scientist says after 10-year analysis of da Vinci masterpiece'. *The New Mail*. Available at: https://www.

thenewmail.co.uk/news/mona-lisa-not-lisa-at-all-scientist-says-after-10-year-analysis-of-da-vinci-masterpiece/ [accessed 15 October 2016].

11. Nikkhah, R. (2015). 'Hidden portrait 'found under Mona Lisa', says French scientist'. *BBC News*. Available at: http://www.bbc.com/news/entertainment-arts-35031997 [accessed 15 October 2016].
12. Ware, D. G. (2015). "Mona Lisa' not 'Lisa' at all, scientist says after 10-year analysis of da Vinci masterpiece'. *United Press International*. Available at: http://www.upi.com/Top_News/World-News/2015/12/08/Mona-Lisa-not-Lisa-at-all-scientist-says-after-10-year-analysis-of-da-Vinci-masterpiece/2031449603461/ [accessed 15 October 2016].
13. Jones, J. (2015). 'There's only one Mona Lisa! Why a 10-year study got it all wrong'. *The Guardian*. Available at: https://www.theguardian.com/artanddesign/jonathanjonesblog/2015/dec/08/leonardo-mona-lisa-why-a-10-year-study-got-it-all-wrong [accessed 15 October 2016].
14. Vincent, A. (2015). 'Mona Lisa discovery: Scientist claims secret, second portrait of 'real' Mona Lisa lies underneath painting'. Available at: http://www.telegraph.co.uk/art/artists/mona-lisa-discovery-scientist-claims-secret-second-portrait-of-r/ [accessed 15 October 2016].
15. Martin Kemp Reply on Mona Lisa by Leonardo da Vinci (2015). Available at: http://www.academia.edu/19624789/Martin_Kemp_Reply_on_Mona_Lisa_by_Leonardo_da_Vinci_december_2015_ [accessed 15 October 2016].
16. Miller, N. (2013). 'And she just keeps on smiling'. *Sydney Morning Herald Entertaining*. Available at: http://www.smh.com.au/entertainment/art-and-design/and-she-just-keeps-on-smiling-20130603-2nkn1.html [accessed 10 May 2014].
17. Reilly, J. and Ellicott, C. (2013). 'Is this the first Mona Lisa? Experts say Leonardo did paint an early version of the iconic masterpiece after months of tests'. *Mail Online*. Available at: http://www.dailymail.co.uk/news/article-2278126/Mona-Lisa-Is-really-Leonardos-prequel-famous-portrait.html [accessed 10 May 2014].
18. Hewitt, S. (2012) "Young Mona'? Doubts grow as experts hostile, evasive or say 'misrepresented"'. *Huffington Post*. Available at: http://www.huffingtonpost.com/simon-hewitt/young-mona-lisa_b_1926690.html [accessed 11 May 2014].
19. Bradley, S. (2013). 'Experts query new proof for 'early' Mona Lisa'. *Swissinfo.ch*. Available at: http://www.swissinfo.ch/eng/experts-query-new-proof-for--early--mona-lisa/34999346 [accessed 11 May 2014].
20. Kemp, M. (2012). 'Leonardo da Vinci. The Isleworth Mona Lisa'. *This and That*. Available at: http://martinkempsthisandthat.blogspot.com/

search?updated-min=2012-01-01T00:00:00-08:00&updated-max=2013-01-01T00:00:00-08:00&max-results=17 [accessed 23 May 2014].

21. Bradley, S. (2012). 'Group unveils "early" Mona Lisa in Geneva'. *Swissinfo. ch*. Available at: https://www.swissinfo.ch/eng/culture/scoop---or-not-_group-unveils--early--mona-lisa-in-geneva/33606794 [accessed 10 February 2014].
22. Benford, S. 'Famous paintings reviewed. Famous paintings: Mona Lisa'. *Masterpiece Cards*. Available at: http://www.themasterpiececards.com/famous-paintings-reviewed/bid/94642/Famous-Paintings-Mona-Lisa [accessed 8 May 2014].
23. Daley, S. (2012). 'Beneath that beguiling smile, seeing what Leonardo Saw'. Available at: http://thehimalayanvoice.blogspot.com/2012/04/beneath-that-beguiling-smile-seeing.html [accessed 9 March 2014].
24. Mozo, A. G. (2014). *Leonardo da Vinci Technical Practice: Paintings, Drawings and Influence*. Menu, M. (ed.), Hermann, p. 194.
25. Mottin, B. (2014). *Leonardo da Vinci Technical Practice: Paintings, Drawings and Influence*. Menu, M. (ed.), Hermann, p. 208.
26. Museo Del Prado (2012). 'The Museo del Prado is presenting the conclusions of the technical study and restoration of its version of La Gioconda'. Available at: https://www.museodelprado.es/en/whats-on/new/the-museo-del-prado-is-presenting-the-conclusions/32085355-9f09-4e16-a516-23e70e383880 [accessed 19 September 2016].
27. Byrne, D. (2010). 'Connoisseurship Today-Art History Today'. *Art History Today*. Available at: http://artintheblood.typepad.com/art_history_today/2010/07/connoisseurship-today.html [accessed 12 September 2014].

PART IV

THE SCIENTIST

1. Johansson, S. A. E. (1986). 'Proton-induced X-ray emission (PIXE) spectrometry: State of the art', *Fresenius' ZeitschriftfüranalytischeChemie*, **324**(7), pp. 635–641.
2. Arrowsmith, P. (1987). 'Laser ablation of solids for elemental analysis by inductively coupled plasma mass spectrometry', *Analytical Chemistry*, **59**(10), pp. 1437–1444.
3. Nothnagle, P. E., Chambers, W. and Davidson, M. W. (2013). 'Introduction to Stereomicroscopy', in Bucklow, S. (1996). *Formal Connoisseurship and the Characterisation of Craquelure*. Ph.D Thesis, University of Cambridge.

4. Varley, A. J. (1999). *Statistical Image Methods for Line Detection*, Ph.D Thesis, University of Cambridge.
5. Matthaes, G. (2000). *The Art Collector's Illustrated Handbook: How to Tell Authentic Antiques From Fakes*, Volume 1. Milano: Museo del Collezionista d'Arte, p. 17.
6. Wieseman, M. E. (2010). *A Closer Look Deception and Discoveries*. National Gallery, London, p. 41.
7. Plester, J. (1956). 'Cross section and chemical analysis of paint samples', *Studies in Conservation*, **2**(3), pp. 110–157.
8. John Innes Center (2014). 'What is light microscopy? Polarised light microscopy'. Available at: https://www.jic.ac.uk/microscopy/intro_LM.html [accessed 20 September 2014].
9. Scherrer, N. C., Stefan, Z., Francoise, D., Annette, F. and Renate, K. (2009). 'Synthetic organic pigments of the 20th and 21st century relevant to artist's paints: Raman spectra reference collection', *Spectrochimica Acta, Part A: Molecular and Biomolecular Spectroscopy*, **73**(3), pp. 505–524.
10. Brostoff, L. B., Centeno, S. A., Ropretm P., Bythrow, P. and Pottier, F. (2009). 'Combined X-ray diffraction and Raman identification of synthetic organic pigments in works of art: From powder samples to artists' paints', *Analytical Chemistry*, **81**(15), pp. 6096–6106.
11. Smith, G. D. and Clark, R. J. H. (2004). 'Raman microscopy in archaeological science', *Journal of Archaeological Science*, **31**(8), pp. 1137–1160.
12. Edwards, H. G. M., Drummond, L. and Russ, J. (1993). 'Fourier transform Raman spectroscopic study of prehistoric rock paintings from the Big Bend region, Texas', *Journal of Raman Spectroscopy*, **30**(6), pp. 421–428.
13. Rull Perez, F., Edwards, H. G. M., Rivas, A. and Drummond, L. (1999). 'Fourier transform Raman spectroscopic characterization of pigments in the mediaeval frescoes at Convento de la Peregrina, Sahagun, Léon, Spain. Part 1 — Preliminary study', *Journal of Raman Spectroscopy*, **30**(4), pp. 301–305.
14. Brown, K.L. and Clark, R. J. H. (2002). 'Analysis of pigmentary materials on the Vinland Map and Tartar Relation by Raman microprobe spectroscopy', *Analytical Chemistry*, **74**(15), pp. 3658–3661.
15. Brown, K. L. and Clark, R. J. H. (2004a). 'Analysis of key Anglo-Saxon manuscripts (8–11th centuries) in the British Library: Pigment identification by Raman microscopy', *Journal of Raman Spectroscopy*, **35**(3), pp. 181–189.
16. Brown, K. L. and Clark, R. J. H. (2004b). 'The Lindisfarne Gospels and two other 8th century Anglo-Saxon/Insular manuscripts: Pigment identification by Raman microscopy', *Journal of Raman Spectroscopy*, **35**(1), pp. 4–12.

17. Kennedy, R. (2005). 'Is this a real Jackson Pollock?' *The New York Times*. Available at: http://www.nytimes.com/2005/05/29/arts/design/29kenn.html? pagewanted=all&_r=0 [accessed 13 July 2014].
18. Derrick, M. R., Stulik, D. C. and Landry, J. M. (1999). *Infrared Spectroscopy in Conservation Science*. Getty Conservation Institute, Los Angeles.
19. Thermo Nicolet (2014). 'Introduction to Fourier Transform Infrared spectrometry'. Available at: https://www.google.com.eg/url?sa=t&rct=j&q=&esrc=s&source=web&cd=1&cad=rja&uact=8&ved=0ahUKEwiYhfOgh4fXAhXSZVAKHbaNACkQFggsMAA&url=https%3A%2F%2Fwww.researchgate.net%2Ffile.PostFileLoader.html%3Fid%3D57ce879ded99e1e0364482dd%26assetKey%3DAS%253A403246750420994%25401473152925890&usg=AOvVaw37zb_muZurKPXoOTDEaYkD [accessed 20 September 2014].
20. Rosi, F., Miliani, C., Braun, R., Harig, R., Sali, D., Brunetti, B. G. and Sgamellotti, A. (2013). 'Noninvasive analysis of paintings by mid-infrared hyperspectral imaging', *Angewandte Chemie*, **125**(20), pp. 5366–5369.
21. Bonaduce, I. and Andreotti, A. (2009). 'Py-GC-MS of organic paint binders', in Colombini, M. P. and Modugno, F. (eds.) *Organic Mass Spectrometry in Art and Archaeology*, John Wiley & Sons Ltd., Chichester, pp. 303–326.
22. Kusch, P. (2012). 'Pyrolysis-gas chromatography/mass spectrometry of polymeric materials', in Mohd M.A. (ed.), *Advanced Gas Chromatography-Progress in Agricultural, Biomedical and Industrial Applications*, pp. 343–362.
23. Russell, J., Singer, B. W., Perry, J. J. and Bacon, A. (2011). 'The identification of synthetic organic pigments in modern paints and modern paintings using pyrolysis-gas chromatography–mass spectrometry'. *Analytical and Bioanalytical Chemistry*, **400**(5), pp. 1473–1491.
24. Evershed, R. P. (2000). 'Bimolecular analyses by mass spectrometry', in Ciliberto, E., and Spoto, G. (eds.), *Modern Analytical Methods in Art and Archaeology*. John Wiley & Sons Ltd., Chichester, pp. 177–239.
25. Smith, K., Horton, K., Watling, R. J. and, Scoullar, N. (2006). 'Detecting art forgeries using LA-ICP-MS incorporating the in situ application of laser-based collection technology'. *Talanta*, **67**(2), pp. 402–413.
26. Hill, S. J. (ed.). (2006). *Inductively Coupled Plasma spectrometry and Its Applications*, Second Edition, John Wiley & Sons Ltd., Chichester.
27. Hou, X. and Bradley, T. J. (2000). 'Inductively Coupled Plasma/Optical Emission Spectrometry', in Meyers, R.A. (ed.), *Encyclopedia of Analytical Chemistry*, John Wiley & Sons Ltd., Chichester, pp. 9468–9485.
28. Varoli, J. (2008). 'Can past nuclear explosions help detect forgeries?' Available at: http://connection.ebscohost.com/c/articles/32643703/can-past-nuclear-explosions-help-detect-forgeries [accessed 4 August 2014].

29. Carltlidge, E. (2008). 'Nuclear fallout used to spot fake art'. Available at: http://physicsworld.com/cws/article/news/2008/jul/04/nuclear-fallout-used-to-spot-fake-art [accessed 4 August 2014].
30. Fabian, D. and Fortunato, G. (2010). 'Tracing white: A study of lead white pigments found in seventeenth-century paintings using high precision lead isotope abundance ratios', in Kirby, J., Nash, S. and Cannon, J. (eds.), *Trade in Artists' Materials: Markets and Commerce in Europe to 1700*, Archetype Publications, London, pp. 426–443.
31. Fortunato, G., Ritter, A. and Fabian, D. (2005). 'Old Masters' lead white pigments: Investigations of paintings from the 16th to the 17th century using high precision lead isotope abundance ratios', *Analyst*, **130**(6), pp. 898–906.
32. Richter, A. (2009). 'Carbon dating & archaeology'. *CyArk*. Available at: http://www.cyark.org/news/carbon-dating-and-archaeology?gclid=CLT-zP6Iv6YCFdNl7AodimjVJA [accessed 2 October 2014].
33. Parkes, P. A. (1986). *Current Scientific Techniques in Archaeology*, Croom Helm Ltd., London.
34. UC Davis. *Accelerator Mass Spectroscopy*. Available at: http://chemwiki.ucdavis.edu/Analytical_Chemistry/Instrumental_Analysis/Mass_Spectrometry/Accelerator_Mass_Spectroscopy [accessed 30 August 2014].
35. Bruker. 'Handheld XRF spectrometry'. Available at: http://www.bruker.com/products/x-ray-diffraction-and-elemental-analysis/handheld-xrf.html [accessed 23 January 2014].
36. Padfield, J., Saunders, D., Cupitt, J. and Atkinson, R. (2002). 'Improvements in the acquisition and processing of X-ray images of paintings'. *National Gallery London, Technical Bulletin*, **23**(1), 62–75.
37. Schreiner, M., Frühmann, B., Jembrih-Simbürger, D. and Linke, R. (2004). 'X-rays in art and archaeology — An overview', *Powder Diffraction*, **19**(1), pp. 3–11.
38. Janssens, K., Alfeld, M., Van der Snickt, G., De Nolf, W., Vanmeert, F., Radepont, M., Monico, L., Dik, J., Cotte, M., Falkenberg, G., Miliani, C. and Brunetti, B. G. (2013). 'The use of synchrotron radiation for the characterization of artists' pigments and paintings', *Annual Review of Analytical Chemistry*, **6**, pp. 399–425.
39. Saunders, D., Billinge, R., Cupitt, J., Atkinson, N. and Liang, H. (2006). 'A new camera for high-resolution infrared imaging of works of art', *Studies in Conservation*, **51**(4), pp. 277–290.
40. Hendriks, E., Johnson, D. H. and Johnson, C. R. Jr. (2010). 'Interpreting canvas weave matches', *Art Matters*, **5**, pp. 53–61.

41. Van Tilborgh, L., Meendendorp, T., Hendriks, E., Johnson, D. H., Johnson, C. R. Jr. and Erdmann, R. G. (2012). 'Weave matching and dating of Van Gogh's paintings: An interdisciplinary approach'. *Burlington Magazine*, **154**, pp. 112–122.
42. NASA (2004). 'Chronology of Thor-Delta development and operations'. Archived from the original, November 1, 2004, refers to unnamed NASA mission.
43. Maxwel, A. F. B. (2003). 'Multispectral imagery'. Available at: www.au.af.mil/au/awc/space/primer/multispectral_imagery.pdf [accessed 14 January 2014].
44. Kemp, M. and Cotte, P. (2010). *La Bella Principessa: The Story of the New Masterpiece by* Leonardo da Vinci, Hodder and Stoughton, p. 98.
45. Elias, M. and Cotte, P. (2008). 'Multispectral camera and radiative transfer equation used to depict Leonardo's *sfumato* in Mona Lisa'. *Applied Optics*, **74**(12), 2146–2154.
46. Cotte, P. (2010). 'Scanning the Mona Lisa with multispectral camera'. Available at: https://www.youtube.com/watch?v=XOBC2qnWFZg [accessed 12 February 2014].
47. Kuniholm, P. I. (2002). 'Dendrochronology (tree-ring dating) of panel paintings', in Taft, S. and Mayer, J. W. (eds.), *The Science of Paintings*, Springer-Verlag, New York, pp. 206–214.
48. Havenith, T. (2012). 'Notes of Nature: Dendrochronology — or the history and uses of tree rings'. Available at: http://notesofnature.blogspot.com/2012/06/dendrochronology-or-history-and-uses-of.html [accessed 12 February 2014].
49. Richter, A. M. (2014). 'Dendrochronology — Real science VS Hollywood forensics. utilizing tree ring dating for archaeological investigation'. Available at: http://archive.cyark.org/dendrochronology-real-science-vs-hollywood-forensics-blog [accessed 14 January 2014].
50. Miles, D. (1997). 'The interpretation, presentation, and use of tree-ring dates'. *Vernacular Architecture*, **28**(1), pp. 40–56. Available at: DOI: http://dx.doi.org/10.1179/030554797786050563 [accessed 23 October 2013].
51. Craddock, P. (2009). 'Scientific investigation of copies, fakes and forgeries'. Available at: http://www.scribd.com/doc/205637013/Scientific-Investigation-of-Copies-Fakes-and-Forgeries [accessed 28 December 2014].
52. Reimers, P. Riederer, J., Goebbels, J. and Kettschau, A. (1989). 'Dendrochronology by means of X-ray computed tomography', in Maniatis, Y. (ed.), *Archaeometry, Proceedings of the 25th International Symposium*, Elsevier Science Publishers BV, Amsterdam, pp. 121–125.

53. Klein, P. (1998). 'Dendrochronological analyses of art objects', in McCrone, W., Chartier, D. R. and Weiss, R. J. (eds.), *Scientific Detection of Fakery in Art, Proceedings of the International Society for Optics and Photonics (SPIE)*, 3315, pp. 21–30. Available at: doi:10.1117/12.308589 [accessed 28 December 2013].

EPILOGUE

1. Eliot, T. S. (1971). 'Little gidding', *Four Quartets*, Faber and Faber, Copyright Esme Valerie Eliot, p. 58.
2. Helmore, E. (2016). 'Art market in 'mania phase' and risks bursting of the bubble, report says'. *The Guardian*. Available at: https://www.theguardian.com/artanddesign/2016/jan/17/art-market-mania-phase-bubble-report [accessed 30 January 2017].
3. Rivetti, E. (2016). 'Sotheby's buys Orion Analytical lab in fight against art fraud'. *The Art Newspaper*. Available at: http://old.theartnewspaper.com/market/sotheby-s-buys-orion-analytical-lab-in-fight-against-art-fraud-/ [accessed 17 February 2017].
4. Authentication in Art (2017). 'Sotheby's announces the acquisition of Orion Analytical and the appointment of James Martin as director of scientific research'. Available at: http://authenticationinart.org/pdf/artmarket/james-martin-sothebys.pdf [accessed 17 February 2017].
5. Pogrebin, R. (2016). 'Sotheby's hires fraud expert to start new research department'. *The New York Times*. Available at: https://www.nytimes.com/2016/12/05/arts/design/sothebys-hires-fraud-expert-james-martin-orion-analytical.html?_r=0 [accessed 17 February 2017].
6. Cascone, S. (2016). 'Expert forgery-spotter James Martin head to Sotheby's scientific research department'. *Artnet News*. Available at: https://news.artnet.com/market/james-martin-sothebys-scientific-research-771905 [accessed 17 February 2017].
7. Pickford, J. (2016). 'Sotheby's creates forensic art unit to detect forgeries'. *Financial Times*. Available at: https://www.ft.com/content/d24323c4-bad2-11e6-8b45-b8b81dd5d080 [accessed 17 February 2017].
8. Bailey, M. (2010). 'Van Meegeren's secret supplies'. Available at: https://www.toncremers.nl/london-uk-van-meegerens-secret-supplies/ [accessed 25 March 2014].
9. Saverwyns, S. and Fremout, W. (2011). 'Genuine or fake: A micro-raman spectroscopy study of an abstract painting attributed to Vasily Kandinsky'. Available at: http://www.ndt.net/article/art2011/papers/SAVERWYNS%20-%20M%208.pdf [accessed 4 September 2014].

10. Eastaugh, N. Director of Research at Art Access & Research, e-mail sent to Jehane Ragai, 14 July 2014.
11. Reagan, G. (2007). 'Scientist reveals Pollocks are fake'. Available at: http://observer.com/2007/11/scientist-reveals-pollocks-are-fake/ [accessed 23 October 2017].
12. Taylor, K. (2007). 'Pollock matters', *Art Around Town. The Saga of the 'Matter Pollocks'*. Available at: http://www.nysun.com/arts/pollock-matters/67324/ [accessed 4 March 2014].
13. Pickford, J. and Spero, J. (2016). 'Old Master market reels from Sotheby's fake assessment'. *Financial Times*. Available at: https://www.ft.com/content/23a4b5ac-8b08-11e6-8cb7-e7ada1d123b1 [accessed 17 February 2017].
14. Alberge, D. (2014). 'Revealed: The art experts who pass fakes as authentic. Dubious data and selective use of evidence lead to forgeries being attributed to masters'. *The Observer*. Available at: http://www.theguardian.com/artanddesign/2014/feb/23/art-scholars-disgrace-forgeries [accessed 2 May 2014].
15. Flynn, T. (2014). 'Oxford Seminar calls for more rigorous standards in technical art history'. Available at: http://tom-flynn.blogspot.com/2014/04/oxford-seminar-calls-for-more-rigorous.html [accessed 2 May 2014].
16. Authentication in Art (2016).'Authentication in Art congress 2016 guidelines'. Available at: http://authenticationinart.org/pdf/newsletter/Release-Education-Guidelines.pdf [accessed 17 February 2017].
17. Authentication in Art (2014). 'Congress poster of The Hague Congress on Authenticity in Art'. Available at: http://www.authenticationinart.org/congress-2014/ [accessed 14 September 2014].
18. Slogett, R. and Kowalski, V. (2014). 'Building evidence for use in criminal cases – standard practice and methodologies – A case study in Australia'. Available at: http://authenticationinart.org/pdf/papers/Building-evidence-for-use-in-criminal-cases-%E2%80%93-standard-practice-and-methodologies-%E2%80%93-A-case-study-in-Australia-Robyn-Sloggett-and-Vanessa-Kowalski.pdf [accessed 14 September 2014].
19. University of Delaware (2016). Department of Art History. Available at: https://www.arthistory.udel.edu/ [accessed 18 February 2016].
20. Wieseman, M. E. (2010). *A Closer Look: Deceptions and Discoveries*, National Gallery, Yale University Press, London, p. 9.
21. Ardid, M., Ferrero, J.L., Juanes, D., Roldán, C., Crespo, M., Pernett, M.E., Marzal, M., Burke, M., Rovira, S. and Vives, R. (2014). 'Identification of forged works of art by portable EDXRF spectrometry'. Available at: www.icdd.com/resources/axa/vol46/v46_56.pdf [accessed 31 October 2014].

22. Joanna, E., Brian, R. S., Singer, W., Perry, J. J. and Bacon, A. (2012). 'The materials and techniques used in the paintings of Francis Bacon (1909–1992)', *Studies in Conservation*, **57**(4), pp. 207–217.
23. Stege, H. (2014). 'Authenticity expertises at the Doerner Institute – A look on current practice and future necessities with focus on organic pigment analyses'. Available at: http://authenticationinart.org/pdf/newsletter/Authentication-in-Art-Newsletter-Christmas-2013.pdf [accessed 14 September 2014].
24. Chartier, D. and Notehelfer, F. (2006). 'Authentication: Science and art at odds', in McCrone, W. (ed.), *Scientific Detection of Forgery in Art*, SPIE press, pp. 3315–3409.
25. Boime, A. and Kossolapov, A. (2003). 'Manet's lost Infanta', *Journal of the American Institute for Conservation*, **42**(3), pp. 407–418.
26. Helmore, E. (2016). 'Art market in 'mania phase' and risks bursting of the bubble, report says'. The Guardian. Available at: https://www.theguardian.com/artanddesign/2016/jan/17/art-market-mania-phase-bubble-report [accessed 30 January 2017].

Picture Credits

Ahram online

Fig. 26: *The Girl with Green Eyes*, 1931 rendition.

Art Roster, Mt Shasta

Fig. 46: Copies, copies, copies.

Badrawi, Fadia

Figs. 2: Han van Meegeren (1889–1947), 4: John Myatt, 8: Sandro Botticelli (1445–1510), 9: Umberto Giunti (1886–1970), 13: Wolfgang Beltracchi (b. 1951–), 25: Mahmoud Said (1897–1964), 34: Elena Basner (b. 1956–), 39: Ann Freedman, 43: How blockchain works for physical art, 53: Different layers in a painting, 54: Principle of an ordinary microscope and microscope with a CCD camera, 55: The polarised light microscope, 56: Polariser and polarised wave, 57: A raster scan, 58: Generation of X-rays and back-scattered electrons, 59a: A conventional Raman spectrometer, 59b: Raman microscope, 60: Typical mass spectrometer, 61: Arrangement for pyrolysis-gas chromatography, 62: Principle of LA–ICP-MS, 64: Formation and decay of ^{14}C, 65: Principle of mass spectroscopy of carbon isotopes, 66: Principle of XRF, 68: Arrangement for radiographic analysis, 69: Production of synchrotron radiation, 70: X-ray diffractometer, 71: Principle of IR reflectography, 72: Weaving in a canvas, 75: Tree rings, 76: The proper approach to authentication and 77: An integrated approach.

Bridgeman Images and Christie's
Fig. 32: *Saint Praxedis* by Johannes Vermeer.
(Delft 1632–1675) signed and dated 'Meer 1655' (lower left) oil on canvas 40 × 32 in (101.6 × 82 cm) Unknown owner.

Cotte, Pascal
Fig. 45: *La Bella Principessa.*
Fig. 48: Multispectral imaging of Leonardo's *Mona Lisa.*
Fig. 73: The multispectral camera's 13-filter wheel.
Fig. 74: Multispectral views in colours of *The Lady with an Ermine* by Leonardo da Vinci.

Courtauld Institute of Art, Somerset House, Strand, London Photo Credit:
Fig. 10: *Madonna of the Veil.*

EPA Sales
Fig. 50: The Isleworth *Mona Lisa.*

Fortuanato, Giuseppe (Swiss Federal Laboratories for Materials Science and Technology)
Fig. 63: Origin of the lead ore from which lead was extracted.

Kimbell Art Museum
Fig. 36: *The Cardsharps.*

Kröller–Müller Museum
Fig. 30: Underdrawing showing the two wrestlers.
Fig. 31: *Sunset at Montmajour.*

Marc Forgy
Fig. 1: Elmyr de Hory (1906–1976).

Mahmoud Shaltout
Fig. 3: John Drew (b.1948–).
Fig. 5: Shaun Greenhalgh (b.1961–).
Fig. 12: Heinrich Campendonk (1889–1957).

Fig. 23: Lucian Freud (1922–2011).
Fig. 37: Domenico de Sole (b.1944–).
Fig. 40: Lawyer interrogating Domenico de Sole.

National Gallery London
Fig. 17: *The Virgin and Child with an Angel.*
Fig. 18: Faked craquelure observed on *The Virgin and Child with an Angel.*
Fig. 19: Reflectogram of the Angel in the National Gallery's painting *The Virgin and Child with an Angel* assumed to have been painted by Francesco Francia (1450–1517).

Norton, Robert
Fig. 44: Certificate of authenticity verified by Verisart.

Philip Mould & Company
Fig. 15: *Nude* — a Chagall forgery.
Fig. 24: *Man in a Black Cravat.*

Prado National Museum
Fig. 52: The Prado *Mona Lisa.*

Public Domain Open Access
Figs. 6: Paul Gauguin (1848–1903), 7: Self engraving of Giovanni Morelli (1816–1891), 16: Marc Chagall (1887–1985), 20: Venus Forgery attributed to Lucas Cranach the Elder, 21: *David contemplating the head of Goliath*, 22: Pollock style 2009 rendition, 27: *The Girl with Green eyes* 1932 rendition, 28: Vincent van Gogh (1853–1890), 33: *Salvator Mundi*, 35: *Paris Café*, 38: *Untitled* 1956, forged Mark Rothko, 41: The rule of law, 42: DNA, 51: Leonardo's *Vitruvian man.*

Rover Images (*Permission Granted for Cover and Interior of Book*)
Fig. 14: *Red Picture with Horses* forgery attributed to Campendonk.

Sipa Press
Fig. 49: Hidden portrait found by Cotte under *Mona Lisa.*

Thomas, John Meurig
Fig. 67: XRF spectrum.

Van Gogh Museum
Fig. 29: *Still Life with Meadow Flowers and Roses.*

Wimmer, Gabriele
Fig. 12: Doctored picture of Helene Beltracchi with some of the forgeries.

Glossary Second Edition

***A Maid Asleep*:** An oil painting by the Dutch artist Johannes Vermeer. Painted in the period 1656–1657. Currently found in the Metropolitan Museum of Art.

Accelerator Mass Spectrometry (AMS): A technique that separates rare isotopes from the more abundant isotopes of an element, e.g. ^{14}C from ^{12}C. It accelerates the ions under analysis to extremely high kinetic energies before analysing their masses.

Adam, Orietta: Friend of art collector Sir Denis Mahon.

Anachronistic: Chronologically inconsistent such as an artwork or a pigment that is not in its correct historical time.

Anisotropic pigments: Pigments exhibiting properties with different values when measured in different directions of light (e.g. refractive index).

Apollinaire, Guillaume (1880–1918): A French poet, playwright, novelist, art critic and one of the forefathers of surrealism. Considered as one of the most important French literary figures of the early twentieth century.

Art historian: A person who studies the historical aspects of artworks, who can provide a chronological description of their development and who is concerned with the political and social conditions surrounding their creation.

Authentication: Confirming the truth of an attribute: it is the process of confirming whether something or someone is what or who it is declared to be.

Avant-garde: Innovative people on works in the art, literary and musical fields.

Back-scattered electrons: When an incident beam of electrons collides with the nucleus of an atom and is sent back approximately in the same direction it came from and with little energy loss.

Bacon, Francis (1909–1992): Irish-born British painter known for his bold, expressive and sometimes grotesque representation of the figure and face suggesting estrangement and suffering.

Beltracchi, Wolfgang (b. 1951): Notorious German art forger and artist who together with his wife has admitted to producing over 300 falsely attributed paintings. The forgeries are entirely newly created works in the style of impressionist and modern great masters.

Bernheimer, Konrad: German art dealer and collector (b. 1950).

Bharara, Preet: American lawyer who served as U.S. Attorney for the Southern District of New York from 2009 to 2017.

Blockchain technology: Distributed database, like a public ledger, that maintains a continuously growing list of ordered records called blocks.

Bomb Peak Effect: A dramatic increase in atmospheric radiocarbon levels due to nuclear explosions and testings. Occurred between 1963 and 1965.

Botticelli, Sandro (1445–1510): Leading Italian painter of the early Renaissance who enjoyed great success during most of his lifetime, but became unfashionable, then forgotten after his death. His work was rediscovered only in the 19th century.

Bragg's Law: The fundamental law of X-ray crystallography, $n\lambda = 2d\sin\theta$, where n is an integer, λ is the wavelength of a beam of X-rays incident at an angle θ, on a crystal in which the lattice planes are separated by a distance d. This helps to identify any solid material.

Braque, Georges (1882–1963): A renowned 20th century French painter whose most important contributions to the history of art were the role he played in the development with Picasso of cubism and his alliance with Fauvism.

Bredius, Abraham (1855–1946): Dutch art collector, curator and one of the most authoritative art-historians who dedicated a good part of his life to the study of Vermeer.

Brinkmann, Bodo: Curator of Old Master Paintings, Kunstmuseum in Basel, Switzerland.

Buckland Abbey: A 700-year-old house in Devon, England. It was once the holy home to Cistercian monks and is presently in the ownership of the National Trust, as part museum and part house.

Bukowski's auction house: A Finnish art auction house located in Sweden.

Buonarroti Simoni, Michelangelo di Lodovico (1475–1564): Italian sculptor, painter, architect, and poet, considered to be one of the greatest artists of the Italian High Renaissance period.

Burlington Magazine, The: A monthly academic journal dealing with fine and decorative arts. Established in 1903, it is the longest running art journal in the English language.

Campendonk, Heinrich (1889–1957): Dutch painter born in Germany. He was prohibited from exhibiting when in 1933 the Nazi regime came to power, his art being labelled as degenerate.

Canvas: Woven fabric on which an oil painting is executed, typically stretched across a wooden frame, replacing wooden panels formerly used.

Carraci, Annibale (1560–1609): Italian painter who was one of the driving forces in the creation of the Baroque style.

Catalogue raisonné: Comprehensive, annotated listing of all the known artworks by an artist either in a particular medium or all media.

Cavarozzi, Bartolomeo (1590–1625): Italian caravaggisti painter of the Baroque period active in Spain, alongside his master Giovanni Battista Crescenzi.

Cesium 137 (^{137}Cs): A radioactive isotope of cesium, a common fission product by the nuclear fission of uranium 235 (^{235}U). It has a half-life of about 30.17 years, and decays into barium 137 (^{137}Ba).

Chagall, Marc (1887–1985): Marc Chagall was a prolific multifaceted Russian-born French painter who had a style of his own that combined elements of cubism, expressionism, symbolism and to a lesser extent Modernist art movements.

Charge-coupled device (CCD): Light sensitive integrated circuit etched onto a silicon surface that displays the data belonging to an image, so that each picture element (pixel) in that image is converted to an electrical charge of an intensity that corresponds to a colour in the spectrum. It is a light detector far more efficient than any photographic film.

Charney, Noah (b. 1979): American novelist and art historian. Founder of the Association for Research into Crimes against Art.

Chartier, Roger (b. 1945): French historian and historiographer who is a Professeur in the Collège de France and Annenberg Visiting Professor of History at the University of Pennsylvania. His work in Early Modern European History was rooted in the tradition of the Annales School.

Chopping, Richard Wasey (1917–2008): British illustrator and author best known for painting the dust jackets of Ian Fleming's James Bond novels starting with *From Russia, with Love* (1957).

***Christ and the Disciples at Emmaus* (produced in 1937):** Most successful forgery by Han van Meegeren, received and authenticated enthusiastically by Bredius, the most important art critic of the time.

***Christ in the House of Martha and Mary*:** Also called *Christ in the House of Mary and Martha*. A painting produced in 1655 by the Dutch painter Johannes Vermeer. Currently in the National Gallery of Scotland in Edinburgh.

Christie's: Auction house for fine arts, with the main headquarters located in London.

Clark McKenzie, Kenneth (1903–1983): Author, broadcaster, renowned art historian and aesthete.

Cobalt blue: An artificially made blue pigment, chemically composed of cobalt oxide-aluminium oxide, or cobalt aluminate, $CoAl_2O_4$, made by sintering finely ground CoO and Al_2O_3 (alumina) at 1200°C.

Computer algorithms: Well-defined procedure and unambiguous instructions (with no room for subjective interpretation) that enable a computer to solve a problem.

***Contraste des formes*:** An analytical cubism representation painting by the French artist Fernand Léger.

Cordia, William: Famous Dutch collector. The Cordia family collection includes works by more than 150 famed artists.

Cotte, Pascal (b. 1958): Engineer, founder of Lumiere Technology and inventor of the first multispectral high-definition camera.

The Courtauld Gallery, UK: Art museum in Somerset House, on the Strand in central London. Endowed initially by its founder Samuel Courtauld with a remarkable collection of Impressionist and Post-Impressionist paintings, the gallery now houses one of Britain's finest art collections.

Cranach the Elder, Lucas (1472–1553): German Renaissance painter and printmaker known for his portraits.

Craquelure: A pattern of fine cracking formed on painting surfaces by ageing.

Cross-section analysis: In a painting, analysis of individual pigment layers in a cross-section.

Crussard, Sylvie: Gauguin expert at the Wildenstein Institute in Paris.

Cubism: The most influential art movement of the early 20th century. It was created in Paris by Pablo Picasso and George Braque and refers to art produced in Paris during the 1910s and through the 1920s. It does not depict objects from one viewpoint, but rather analyses and reassembles them in an abstract form to represent multiple viewpoints of the same object.

Cusping: When two curves meet, the pointed end is the 'cusp'.

da Caravaggio, Michelangelo Merisi (1571–1610): Renowned Italian painter active in Rome, Naples, Malta, and Sicily between 1592 and 1610.

da Vinci, Leonardo (1452–1519): Leading artist and intellectual of the Italian Renaissance (as a painter, architect, sculptor, mathematician, inventor…).

da Montefeltro, Federico (1422–1482): Also known as Federico III da Montefeltro, he was one of the most successful warlords of the Italian Renaissance, and lord of Urbino. da Montefeltro constructed one of the greatest libraries in Italy.

Daley, Michael: Sculptor, journalist and illustrator. Present UK director of Artwatch organisation for the protection of the integrity of art works and architecture.

Davide, Gasparotto: Senior curator of paintings based at the Getty Center in Brentwood since 2014. Gasparotto was director of the Galleria Estense museum in Modena, Italy, for the last two years, and spent 12 years as a curator and art historian at the National Gallery of Parma.

de Corradis, Bernardina: Mistress of Lodovico il Moro Sforza and the mother of his two illegitimate daughters Maddalena and Bianca Sforza.

***De Divina Proportione (The Divine Proportions)*:** Book on mathematics and artistic proportions written around 1497 by Luca Pacioli in Milan and illustrated by Leonardo da Vinci. The book was first printed in 1509.

de Hory, Elmyr (1906–1976): Elmyr de Hory was a Hungarian-born painter and notorious art forger who sold more than a thousand fakes to galleries of good repute. His forgeries included works purported to be by Renoir, Picasso, Matisse and Modigliani.

de Kooning, Willem (1904–1997): Dutch–American expressionist painter and abstract expressionist painter, born in Rotterdam, the Netherlands. His most notable work is a series of six paintings with a woman as a central theme.

de Sole, Domenico (b. 1944): De Sole was elected Chairman of Sotheby's Board of Directors in March 2015. He is an Italian businessman and Chairman of Tom Ford International since it was founded in 2005. Sole was the former president and CEO of Gucci Group.

Dendrochronology: Dendrochronology refers to tree-ring dating, which is a scientific method of dating based on the study of annual growth rings of long-lived trees.

***Diana and her Companions*:** A painting by Dutch artist Johannes Vermeer completed in the early to mid-1650s; currently at the Mauritshuis Museum in The Hague.

Diebenkorn, Gretchen: Richard Diebenkorn's daughter.

Diebenkorn, Richard (1922–1993): One of the great American post-war painters. His work is associated with abstract expressionism and the Bay Area Figurative Movement of the 1950s and 1960s.

Digital detector: Detectors that convert collected electromagnetic radiation into pixels (picture elements) to be stored as a digital image.

DNA (Deoxyribonucleic acid): Hereditary material in humans and almost all other organisms. Nearly every cell in a person's body has the same DNA. This molecule carries the genetic instructions used in the growth, development, functioning and reproduction of all known living organisms.

Doerner Institute: Institute founded in 1937 in Munich for the testing and conservation of artistic and historic works.

Dominican Friar: The Dominican Friars are a male religious order within the Catholic Church. They are called Dominicans after their founder, St Dominic, who established the order in 1216.

Drake, Francis (Sir) (1540–1596): English admiral and sea captain, the most renowned seaman of the Elizabethan age, who carried out the second circumnavigation of the globe in a single expedition from 1577 to 1580.

Drewe, John (b. 1948): A British purvey or purveyor of art forgeries who convinced and commissioned artist John Myatt to create them, earning him about £1.8 million.

Dzerzhinsky District: A former district of the federal city of Saint Petersburg merged into newly created Tsentralny District in March 1994.

Eastaugh, Nicholas: Director of Research at Art Analysis & Research (UK) Ltd. and honorary fellow at The University of Oxford.

EDXRF and EDX: EDXRF (Energy-dispersive X-ray Fluorescence) is a simple and accurate technique for the determination of the chemical composition of pigments through the interaction of X-rays with the sample, with subsequent emission of an X-ray spectrum specific of the type of atoms present in the sample. All the elements are excited simultaneously and subsequently separated and identified individually. **EDX** works on the same principle as EDXRF but is generally used in conjunction with a scanning electron microscope.

Electromagnetic spectrum: Term used by scientists to describe the complete range of the frequencies of light from radiowaves to gamma rays via visible light.

Elias, Mady: Researcher and Director of the Art & Optics Laboratory at the CNRS (National Scientific Research Center) in Boucicaut, also professor at the Nanosciences institute in Paris (INSP) and at the University of Pierre and Marie Curie.

Eliot, T.S. (1888–1965): British essayist, publisher, playwright, literary and social critic, and one of the twentieth century's major poets. He was awarded the Nobel Prize in Literature in 1948, "for his outstanding, pioneer contribution to present-day poetry".

Ernst, Max (1891–1976): German-born painter, sculptor, printmaker and poet. One of the pioneers of the Dada and Surrealist movements. He moved to Paris in 1922 and became a close friend of Eluard and Breton. His dreamlike imagery often mocked social conventions.

Fake or Fortune: A BBC television series presented by journalist Fiona Bruce and art dealer and arthistorian Philip Mould. The series examines the provenance or attribution of notable artworks with the expert collaboration of Dr Bendor Grosvenor, a British art dealer and art historian who is known to have discovered a number of important lost works by Old Master artists.

Feaver, William: Former art critic of *The Observer*, and longtime friend of Lucian Freud, who published in 2007 a biography of Freud.

Federal Institute of Technology in Zurich (ETHZ): Ranked as one of the top universities in continental Europe, ETHZ is a science, engineering, mathematics and management university in Zurich, Switzerland.

Feigen, Richard: High-profile New York art dealer.

Feldman, David: Professional philatelist, auctioneer and author.

Ficherelli, Felice (1605–1660): Italian painter belonging to the Baroque period.

Fitzwilliam Museum: Located in central Cambridge (Trumpington Street), the Fitzwilliam Museum is the antiquities and art museum of the University of Cambridge.

Flemish Renaissance school: Refers to the works of artists who produced Renaissance paintings during the 15th and 16th centuries in the Southern Netherlands (today Belgium).

Fluorescence: Emission of electromagnetic radiation (mostly visible light) by a substance as a result of absorbing an incident radiation.

Flynn, Tom: London-based journalist, writer and art-historian.

Forensic scientist: One who helps in investigating crimes through gathering and analysing documents and physical evidence from crime scenes.

Forgy, Mark (b. 1959): Close friend of the late forger Elmyr de Hory.

Fourier Transform Infrared Microscopy (FTIR)/Micro-Fourier Transform Infrared Spectroscopy (micro-FTIR): A technique which is used to obtain an infrared spectrum of a sample of organic material and certain inorganic materials, over a wide spectral range, thereby providing specific information about chemical bonding and molecular structure. Now the preferred method of infrared spectroscopy.

Francia, Francesco (1447–1517): Italian painter, goldsmith and medallist. Many of his paintings were influenced by Perugino and Raphael.

Freud, Lucian (1922–2011): British painter and draftsman, specialising in figurative art, and is known as one of the foremost 20th-century portraitists. He was born in Berlin, the son of a Jewish architect and the grandson of Sigmund Freud. Freud painted mainly those closest to him: friends and family, wives, mistresses, and also himself.

Freud, Sigmund: Austrian neurologist and one of the most influential scientists in the fields of psychology and psychiatry. He is the founder of psychoanalysis, a clinical method for treating psychopathology through dialogue between a patient and a psychoanalyst.

Friedlander, Max (1867–1958): An art historian and a German curator specializing in early Netherlandish painting and the Northern Renaissance. Had he not been friendly with Goering, he would have been destined to a concentration camp after being arrested by the Nazis when he fled in 1939 from Germany to Amsterdam. After the war, he dedicated his life to publishing and writing.

Fry, Eliot Robert (1866–1934): English painter and critic. Establishing his reputation as a scholar of the Old Masters, he became an advocate of more recent developments in French painting, to which he gave the name Post-Impressionism.

Futurism: Important avant-garde artistic and social movement that emerged in Italy by the beginning of the 20th century. Committed to novelty, its members strived to destroy older forms of culture and celebrated advanced technology and urban modernity.

Ganz, Kate: Manhattan gallery dealer who owns the Ganz Gallery.

Gas Chromatography (GC): A technique for separating different substances used for the analysis of compounds after they become vaporised. It is also used to test the purity of substances, as well as to identify compounds.

Gauguin, Paul (1848–1903): French post-impressionist, pioneered a novel style of painting generally identified as Symbolism. He exiled himself in the South Seas where he created a new kind of "primitive" art.

Gautier, Théophile (1811–1872): French poet, dramatist, novelist, journalist, art and literary critic.

Geddo, Cristina: Doctor in History of Art and specialist in the Followers of Leonardo da Vinci.

Gentileschi, Artemisia (1593–1653): Italian Baroque painter, considered one of the most accomplished painters in the generation following that of Caravaggio.

Gerson, Horst (1907–1978): A German-Dutch art historian and Rembrandt scholar and an authority on Netherlandish art. Assisted Abraham Bredius with his Rembrandt catalogue raisonné, and later wrote his own revision in 1968 which reduced the number of attributed Rembrandt works from 639 to less than 419. The Rembrandt Research Project headed by Ernst van de Wetering has looked through the Gerson papers and re-attributed some paintings to Rembrandt.

Gherardini, Lisa: Lisa del Giocondo, also known as Lisa Gherardini. In 2005, Lisa was definitively identified as the model for Leonardo's *Mona Lisa*.

Giacommetti, Alberto (1901–1966): Swiss sculptor, painter and draughtsman. Starting 1947, figures which were very tall and thin characterized his style. He was awarded the First Prize for Sculpture at the Pittsburgh International in 1961, the main prize for sculpture at the Venice Biennale 1962, and the Guggenheim International Award for Painting in 1964.

***Girl with Green Eyes*:** A painting produced in 1931 by the Egyptian artist Mahmoud Said.

Giunti, Umberto (1886–1970): Was from Siena, Italy and is known as the fresco forger, as he painted artwork used for fresco forgery.

Gouache: Similar to watercolour paint. It is opaque rather than transparent, with gum arabic or dextrin as the binder.

Goldman Sachs: The Goldman Sachs Group, Inc. is a leading American global investment banking, securities and investment management firm.

González, Ana: Researcher in the Prado Museum's technical documentation department.

Gouncharova, Natalia (1881–1962): Prominent Russian avant-garde artist who belonged to the circle of Wassily Kandisky. She was a painter, a costume designer and a set designer who expanded the boundaries of avant-garde to cubo-futurism.

Grabar All-Russian Art, Scientific and Restoration Center: Leading Russian institution in Moscow for the scientific study and restoration of artworks.

Graham-Dixon: Born in London in 1960, Andrew Graham-Dixon is one of the leading art critics and presenters of arts television in the English-speaking world.

Greenhalgh, Shaun (b. 1961): British art forger who, over a seventeen-year period, between 1989 and 2006, produced a large number of forgeries in a variety of styles.

He admitted fraud and money laundering at Bolton Crown Court in 2007, and was sentenced to four years and eight months in prison.

Gregori, Mina: Art historian and Professor Emerita at the University of Florence. President of the Foundation of the University of Art History Roberto Longhi in Florence.

Gregoryev, Boris (1886–1939): Famous Russian painter and graphic artist.

Grosvenar, Bendor (b. 1977): British art dealer and art historian. Discovered a number of important lost paintings by Old Masters.

Growth rings: As a tree grows, new layers of wood are formed around the trunk. When the tree is cut down, each layer, visible in a cross section, has two colours, one that is light coloured and grows in the summer and a darker one from the winter. The layers appear as a set of concentric circles and are known as annual rings as their growth period is one year.

Guggenheim Foundation: An institution founded in 1937 in New York by philanthropist Solomon R. Guggenheim and artist Hilla von Rebay, that is concerned with collecting, preserving, and researching modern and contemporary art.

Guggenheim Museum: Established in New York City by the Guggenheim Foundation in 1939 and initially named the Museum of Non-Objective Painting. It was then named The Guggenheim Museum in 1952.

Guggenheim, Peggy (1898–1979): American art collector.

Hals, Frans (1582–1666): Dutch Golden Age-renowned portrait painter. Hals was born in Antwerp, but worked for most of his life in Haarlem, North Holland. He played an important role in the evolution of 17th-century group portraiture and is best known for portraits of the citizens of Haarlem.

Hamilton Kerr Institute: This Institute is a department of the Fitzwilliam Museum, University of Cambridge. Dedicated to the study and conservation of easel paintings and for the restoration of paintings for the Fitzwilliam Museum as well as for public and privately owned collections.

Harvard University Art Museum: Consists of three establishments, the Fogg Museum, the Busch-Reisinger Museum and the Arthur M. Sackler Museum, and four research centres of which one is the Harvard Research Center for the Technical Study of Modern Art.

Harvard University's Center for the Technical Study of Modern Art: This centre was established in 2001 for the study of materials and issues related to the making and conservation of modern works of art.

Hebborn, Eric (1934–1996): British painter and well known forger. He created works that belonged to the Renaissance and Baroque periods. In 1996, he was murdered on the street in Rome.

Hermitage Museum: Founded in 1764 by Catherine the Great, the Hermitage Museum in St Petersburg is one of the oldest and largest museums in the world. It has a strong collection of Italian Renaissance and French Impressionist paintings and an outstanding collection of paintings by Rembrandt, Picasso and Matisse. Its collections comprise over three million items.

Hewitt, Simon: Journalist, reported on art and the art market since 1985 for a welter of international publications, including *Art & Auction, The Art Newspaper, Antiques Trade Gazette* and *Vedemosti* (Moscow).

High-performance Liquid Chromatography (HPLC): Initially referred to as high-pressure liquid chromatography, HPLC is a separation technique used in analytical chemistry to separate, identify and quantify the components of a mixture.

Hiroshima and Nagasaki bombing: In August 1945, the United States dropped atomic bombs on the Japanese cities of Hiroshima and Nagasaki.

Icilio Frederico, Joni (1866–1946): Italian artist known for figure painting and art forgery.

Impasto: A technique of painting whereby very thick paint is laid on the canvas. The paint is usually sufficiently thick so as to reveal the paint-knife or brush strokes.

Impressionism: 19th-century art movement that originated with a group of Paris-based artists. The name Impressionism derives from Monet's 1873 painting *Impression, Sunrise.*

In situ: A Latin phrase literally meaning on site or on the premises.

Inductively coupled plasma (ICP) and Inductively coupled plasma mass spectrometry (ICP-MS): ICP is a source of plasma in which electric currents produced by electromagnetic induction supply the energy. When combined with a mass spectrometer (ICP-MS), it can determine elements at extremely low concentrations.

***Infanta Margarita*:** Early 1860 painting by Manet. Privately owned.

Infrared radiation: Electromagnetic radiation of longer wavelength than visible light and invisible to the human eye. Infrared radiation is given off by all warm objects; it warms all objects it strikes.

Infrared reflectographs: Infrared reflectography images.

Infrared Reflectography: A technique that utilises a wide range of infrared wavelengths that penetrate the paint layers in a painting, revealing details unseen by the human eye. It reveals underdrawings and is particularly helpful in the case of charcoal underdrawings.

Internet of things (IoT): The inter-networking of physical devices over the internet. An example are *smart meters* that allow you to turn down the heating temperature in your home on a sunny day or to turn it off if there is no one at home.

Ion: An atom or molecule with a net charge, either positive or negative, due to unequal total number of electrons and protons.

Isbouts, Jean-Pierre: A historian, bestselling author, and award-winning screenwriter and film director.

Isotopes: Atoms of the same chemical element that have non-identical masses due to different number of neutrons. They may exist either in a stable form, or a radioactive one.

Jagers Collection: Fake art collection of 20th century painters claimed to belong to the German art collector Werner Jagers. The collection idea was invented by Werner's granddaughters to provide a provenance for Beltracchi's forged art works.

Johnson, Barbara Piasecka: A Polish-born American, avid art collector and philanthropist.

Jones, Jonathan: British art critic who has written for *The Guardian* since 1999.

Kandisky, Vasily (1866–1944): Renowned Russian painter who is generally credited to have painted one of the first purely abstract works.

Kemp, Martin: Fellow of the British Academy, Emeritus Professor of the history of art at the University of Oxford. One of the world's leading experts on the art of Leonardo da Vinci.

KGB: Acronym for *Komitet gosudarstvennoy bezopasnosti* was the main security agency for the Soviet Union from 1954 until its break-up in 1991.

Kimbriel, Christine Slottvedd: Paintings conservator at the Hamilton-Kerr Institute, Cambridge, UK.

Kinetic energy: The energy possessed by an object in a state of motion.

Kirkman, James: Art dealer to Lucian Freud for 20 years.

Kitson. Michael (1926–1998): Art historian who became an international authority on the work of the painter Claude Lorrain.

Knoedler Gallery: Renowned art gallery founded in 1846 in New York City. Closed in 2011 after being accused of knowingly selling forged works of arts for tens of millions of dollars.

Koepplin, Dieter: Swiss art historian and curator of the New Digital Archive Museum.

Koetser, Leonard: Art expert and dealer in Flemish Dutch and Italian Old Masters. Owner of the Koetser Gallery.

Kournikova Gallery: Art gallery in Moscow.

Kröller-Müller Museum: An art museum and sculpture garden, founded by art collector Helene Kröller-Müller in 1935 and opened in 1938. It is located in the HageVeluwe National Park in Otterlo in the Netherlands and has the second-largest collection of paintings by Vincent van Gogh, after the Van Gogh Museum.

Krusanov, Andrey: Freelance Russian geochemist, art historian and author of "Russian avante-garde 1907–1932".

Lady Samuel of Wych Cross: Wife of Harold Samuel, Lord of Wych Cross, who collected a great number of paintings during his lifetime.

Landau, Ellen: American art historian, expert on Jackson Pollock.

Landesman, Peter: Former investigative journalist for the New York Times Magazine, novelist and writer who became a film director and producer.

Landis, Mark (b. 1955): American painter and forger, who posing as a philanthropist, donated a large number of forgeries to institutions such as Smithsonian National Portrait and the Art Institute of Chicago as well as to American Art museums. He was exposed in 2008.

Laser Ablation-Inductively Coupled Plasma-Mass Spectrometry (LA-ICP-MS): An elemental and isotopic analytical technique that can be performed directly on solid samples involving the direct conversion of the surface of a sample into an aerosol of fine particles which are carried by helium and ionised in a plasma torch and led to a mass spectrometer.

Laser Amplification Method LAM: The result is like an onion with all the superimposed layers revealed one at a time.

Layer Amplification Method LAM: Technique pioneered by Cotte, P. and Dupouy, M. that allows the reconstruction and analysis of what has taken place within the layers of a painting.

Léger, Fernand (1881–1955): French painter and sculptor, who introduced a special type of cubism. Some of the critics referred to it as 'Tubism' for its emphasis on cylindrical forms.

Lippmann, Gabriel (1845–1921): French physicist who received the Nobel Prize for Physics in 1908 for his method of reproducing colours photographically.

Lomazzo, Paolo (1538–1592): Italian painter, best remembered for his writings on art theory, belonging to the generation that produced Mannerism in Italian art and architecture.

Louvre Museum: One of the largest museums in the world, located on the right bank of the Seine in Paris.

Manet, Edouard (1832–1883): French painter who played a crucial role in the transition from realism to impressionism.

Mapping of the thread density: Paintings are X-rayed revealing the thread patterns. The obtained images when fed into a computer allow the calculation of the individual weave densities as a plot of the average thread count of the vertically or horizontally oriented threads. Such "weave maps" indicate whether paintings originated from the same roll of canvas.

Marchig, Giannino (1897–1983): Italian visual artist.

Marchig, Jeanne: Wife of Giannino Marchig who will be remembered as the ageing animal-lover who took Christie's to court over a Leonardo (*La Bella Principessa*); she died in 2013.

***Martha and Mary Magdalene* (c. 1598):** Painting by the Italian Baroque master Michelangelo Merisi da Caravaggio.

Martin, James: Founder and operator of the Orion Analytical, LLC firm, specializing in the analysis of a wide spectrum of materials including art and artefacts.

Martin, Steve (b. 1945): American actor, comedian, writer, producer and musician.

Mass spectrograph: A mass spectrum is usually a photograph of the intensity vs m/z (mass to charge ratio) plot representing the distribution of ions by mass.

Mass spectrometry (MS): An analytical technique that measures the mass-to-charge ratios as well as the abundance of ions in the gas phase.

Matisse, Henri (1869–1954): French artist who worked in all media, from sculpture to printmaking to painting. Known primarily for his painting. His revolutionary use of bright colours made him one of the most important and influential artists of the 20th century.

Matter, Alex: American filmmaker. The son of Herbert Matter (see Herbert Matter).

Matter, Herbert (1907–1984): Swiss-born, American photographer and graphic designer, whose innovations included the application of photomontage (cutting and joining different photographs into one image) in commercial art.

McCrone, Walter (1916–2002): American chemist who was often referred to as the 'father of modern microscopy' because of his seminal work in the use of light microscopy (particularly polarised light microscopy) in the analysis of materials.

Meedendorp, Teio (b. 1961): Co-author of *On the Verge of Insanity: van Gogh and his Illness* (2016) and a senior researcher at the Van Gogh Museum.

Meyer, Leonard B. (1918–2007): Composer, author and philosopher.

Miro, Joan (1893–1983): Catalan Spanish painter, sculptor and ceramicist, born in Barcelona. Important figure in the history of abstraction. A museum dedicated to his work, the Fundació Joan Miró, was established in his native city of Barcelona.

Modestini, Dianne: A conservator of paintings at the Conservation Centre of the Institute of Fine Arts in New York.

Molecular weight: The mass of a molecule, composed of the sum of all the masses of the constituent atoms in the molecular formula.

***Mona Lisa*:** A half-length portrait of a woman by the Italian artist Leonardo da Vinci, created between 1503 and 1517 in the Renaissance period. It has been in the Louvre since 1797.

Mona Lisa Foundation: A Zurich-based non-profit organization, established with the main goal of analysing and researching what is assumed to be an earlier version of Leonardo's famous *Mona Lisa* painting.

Morelli, Giovanni (1816–1891): Italian art historian and critic who held a degree in medicine.

Morellian analytical approach: It is an empirically based method for the authentication of artworks that was developed in the late nineteenth century by Giovanni Morelli. It is based on identifying small stylistic details that distinguish individual artists.

Mottin, Bruno: Head conservator at the Centre de Recherche et de Restauration des Musées de France, author with Jean-Pierre Mohen and Michel Menu of *Mona Lisa: Inside the Painting*.

Mould, Philip (b. 1960): Philip Jonathan Clifford Mould is an English art dealer and art-historian who specialises in British portraits. He is also a writer on art, a broadcaster and journalist.

Mustad, Christian: Christian Nicolai Mustad, a Norwegian industrialist and art collector who purchased van Gogh's painting *Sunset at Montmajour* in 1908.

Myatt, John (b. 1945): British artist and reformed art fraudster who together with John Drewe sold more than 200 forgeries purporting to be of famous 19th and 20th century painters. He revealed how, using acrylic and mixing it with KY jelly, he created his forgeries. Today, he sells legitimate fakes to art lovers, oligarchs and celebrities.

National Gallery London: Art museum established in 1824 in London, and containing over 2,000 paintings belonging to the period from the mid-13th century to the 18th century.

Nazis: Members of the National Socialist German Workers' Party, which controlled Germany. Under Adolf Hitler, the party advocated a totalitarian government, anti-Semitism and Aryan supremacy.

New York Assembly: Lower house of the New York State Legislature. The Assembly is composed of 150 members representing an equal number of districts, with each district having an average population of 128,652. Assembly members serve two-year terms without term limits.

New York City Bar Association: Consisting of 24,000 lawyers this association is dedicated to improving the administration of justice and promoting the study of law.

Noce, Vincent: Journalist at *The Art Newspaper*.

Nonvolatility: Not liable to evaporate rapidly.

Notehelfer, Fred G.: Professor and since 1992 director of UCLA (University of California, Los Angeles) Center for Japanese Studies.

***Novum Organon*:** Philosophical work by Francis Bacon, written in Latin and published in 1620. It challenged Bacon's contemporaries to discard pure deductive reasoning and embrace a scientific method based on observation and experience.

Organic compounds: Chemical compounds composed mainly of carbon and hydrogen. Other atoms, such as oxygen, sulphur, phosphorus or halogens, may also be present.

Organic pigment: Chemical compounds used as colorants containing mainly carbon chains and carbon rings. They may include also stabilising inorganic elements.

Pacioli, Luca (1445–1517): Referred to as the 'father of accounting and bookkeeping,' Pacioli was an Italian mathematician who collaborated with Leonardo da Vinci and published in 1494 the important book entitled *Summa,* which gave a summary of all the mathematics of the time.

Palette: A surface generally made out of wood used by painters to mix pigments. The term is also used to refer to the range of pigments (colour as well as chemical composition) characteristic of each painter.

Panofsky, Erwin (1892–1968): German art historian whose academic career was pursued in the US after the rise of the Nazi regime.

Parmigianino: Girolamo Francesco Maria Mazzola, known throughout his artistic career as Parmigianino (a nickname meaning "the little one from Parma"), was one of the Italian Renaissance's great geniuses. He was active in Florence, Rome, Bologna and his native city of Parma.

Pastiche: Work of visual art, literature, theatre or music that imitates another work, artist or period.

Pentimenti: *Originates from the Italian word for repentance pentire meaning to repent.* An alteration in a painting identified through the reappearance of earlier images or details that have been modified and painted over by the artist.

Perestroika: A political movement, literally meaning "restructuring", introduced in the mid 1980s to restructure the Soviet political and economic systems. It sought to bring the Soviet Union up to economic par with capitalist countries like Japan, the United States and Germany. This reform policy of openness is viewed as the cause of the dissolution of the Soviet Union and the end of the cold war.

Photosynthesis: A process by which green plants convert light energy (from the sun) into chemical energy in the form of carbohydrates.

Phthalocyanines: Intensely blue-green coloured organic compound used extensively in dyeing.

Picasso, Pablo (1881–1973): One of the greatest and most influential artists of the 20th century; co-founder of the Cubist movement.

Pigment 254: Known as Ferrari red or diketopyrrolopyrrole (DPP red), it is an artificially made organic pigment of molecular formula $C_{18}H_{10}C_{12}N_2O_2$ that was first developed in 1974 at Michigan State University by professor Donald G. Farnum. In 1983, Ciba-Geigy Ltd., now Ciba Specialty Chemicals, based in Basel, Switzerland, patented the first known method of producing it and revolutionized the automobile paint industry.

Pigments: Chemical compounds, including some minerals and synthetic compounds used as colourants, which transmit only selected wavelengths of visible light, thus making them appear colourful.

Pixels: Word originating from "picture element" and is one of the tiny dots that make up a graphic image in a computer's memory.

Plasma: Ionized gas in which electrons and ions coexist in an overall electrically neutral medium. It is considered to be a fourth state of matter.

Pollock, Jackson (1912–1956): Renowned American painter and the most prominent figure behind the abstract expressionist movement in the art world.

Polymer: A large molecule composed of many repeated (monomer) subunits.

Prado Museum: Located in central Madrid and known as the Museo del Prado, it is the main Spanish national art museum. It contains fine collections of European art, dating from the 12th to the early 19th century.

Predetti, Carlo: Armand Hammer Professor Emeritus of Leonardo Studies at UCLA (University of California, Los Angeles). Arthistorian and author of more than 50 books and 700 articles and essays on Leonardo da Vinci.

Princeton University Art Museum: Princeton University's gallery of art, founded in 1882, and located in Princeton, New Jersey. Its permanent collections range from antiquities to contemporary art and focus primarily on works by American, European and Latin American artists. Now houses over 72,000 works of art.

Proton-induced or particle-induced X-ray emission (PIXE): A non-destructive elemental analysis technique that determines the elemental composition of a sample, through exposure to a proton beam that causes atomic interactions. These give off X-rays specific to the component elements. It is now used routinely by art conservators to address questions of provenance and authenticity.

Prussian blue: First modern artificial blue pigment discovered in 1704, chemically it is *Iron(III)-hexacyanoferrate(II).*

Pyrolysis: Thermal decomposition of materials under vacuum or in an inert atmosphere.

Pyrolysis–gas chromatography–mass spectrometry (Py-GC-MS): An analytical technique in which the sample is heated to 600–1000°C, causing large molecules to break down into smaller, more volatile fragments that can be separated by gas chromatography. The data can be used either to identify the material or to obtain some structural information.

Queen Farida (1921–1988): Born Safinaz Zulficar, she was the first wife of King Farouk of Egypt from 1938 to 1948. The marriage was dissolved mainly because she did not give Farouk a male heir to the throne. She was the niece of artist Mahmoud Said.

Radiocarbon dating: A method for determining the age of an object, usually not older than 60,000 years, based on the measurement of its radiocarbon (^{14}C) content.

Raking light: Illuminating objects with a light source at an oblique angle to the surface revealing more clearly the surface texture.

Raman microscopy: Very useful technique for the identification of individual components in pigment mixtures. It uses a Raman microscope specially designed for Raman spectroscopy.

Raman spectroscopy: Raman spectroscopy is a technique commonly used in chemistry. Typically, a sample is illuminated with a low powered laser beam and the object under investigation scatters characteristic wavelengths that act as fingerprints for the identification of pigments.

Refractive index: Ratio between the speed of light in vacuum and its speed in a specific medium.

Rembrandt, Harmenszoon van Rijn (1606–1669): Dutch Baroque artist and etcher, considered as one of the greatest painters and printmakers in European history of art and one of the most important painters in Dutch history.

Renaissance: Comes from French, meaning "rebirth" and represents a cultural movement following the Middle Ages and spanning the period approximately for the 14th to the 17th century. It began in Italy and later spread to the rest of Europe and was characterised by a revival of classical learning and values following a period of stagnation and cultural decline.

Renoir, Pierre-Auguste (1841–1919): Painter, considered one of the leaders of the impressionist art movement.

Retention time: The interval between the instant of introducing a sample, and the detection of its components.

Rijksmuseum: Main national museum in the Netherlands dedicated to the arts and history, spanning the time from the Middle Ages to the present.

Robert, Simon: Consultant, appraiser, and private dealer; President, Robert Simon Fine Art, Inc.

Rose, Vivien (b. 1960): Dame Vivien Judith Rose DBE is a British judge of the High Court of Justice.

Rothko, Mark (1903–1970): Born Markus Yakovlevich Rothkowitz, Rothko was an American painter of Russian Jewish descent. He is generally identified as an abstract expressionist and is considered, with Jackson Pollock and Willem de Kooning, as one of the most famous post-war American artists. His most important works included large-scale paintings of bright coloured rectangles, aimed at evoking emotional responses.

Rubino, Alfonso: Specialist in the geometry of Leonardo.

Russian Federal Agency for Culture: Federal agency investigating organisations or museums using state funds of the Russian Federation. It is a subsection of the Ministry of Culture.

Russian Grabar Scientific Restoration Centre: Centre for art restoration located in Moscow.

Said, Mahmoud (1897–1964): Egyptian judge and renowned modern painter.

***Saint Praxedis*:** An oil painting attributed to the Dutch painter Johannes Vermeer.

Sakhai, Ely (b. 1952): An American civil engineer and art dealer who owned Lower Manhattan art galleries and who was sentenced to 41 months in a federal prison for selling forged art. He continues, after his release, to operate The Art Collection in Great Neck, New York.

Salvator, Mundi: Devotional oil painting of Christ as Salvator Mundi, or Savior of the World recently attributed to Leonardo da Vinci and created in the period 1490–1519. This painting was lost, rediscovered, restored and exhibited in 2011.

Savonarola, Girolamo (1452–1498): Italian Dominican friar and preacher active in Renaissance Florence. He was known for his prophecies of civic glory, the destruction of secular art and culture and his calls for Christian renewal. Savonarola wanted Florence to be a Christian republic with God as governor.

Scanning Electron Microscopy (SEM): A microscope that scans the sample with a beam of electrons to produce an image that gives information about the sample's surface topography. The results are analysed and the composition is determined in conjunction with an energy-dispersive spectrometer (EDXRF).

Scully, Sean (b. 1945): Irish-born American-based painter and printmaker. His work is collected in major museums worldwide and he has twice been nominated for the Turner Prize.

Sforza, Ludovico (1452–1508): Duke of Milan from 1494 until 1499, following the death of his nephew Gian Galeazzo Sforza.

Sfumato: One of the painting modes of the Renaissance characterised by blurred outlines, allowing areas to blend into one another. *Sfumato* means "smoky" in Italian.

Sheldon, Libby: Painting Analyst who founded and runs at UCL the Painting Analysis and Research Unit.

Shishkin, Ivan Ivanovich (1832–1898): Russian landscape painter, associated with the Peredvizhniki (The Wanderers) movement that protested academic restrictions in artistic expression. The Peredvizhniki movement also criticized the inequities and injustices in social life.

Silverman, Peter: Canadian collector and art adviser.

Smith, GD: Analytical chemist from Duke University who completed postdoctoral research at the National Gallery of Art in Washington, DC, the National Synchrotron Light Source at Brookhaven National Laboratory in New York, and at University College London.

***Soleil couchant à Arles* or *Sunset at Arles*:** An oil painting created in 1888 by the Dutch painter Vincent van Gogh.

Sorolla, Joaquín (1863–1923): Renowned Spanish painter who excelled in portraits and landscapes and had a distinct ability to depict the effects of light.

Sotheby's: A multinational auction house corporation, originally founded in London, currently headquartered in New York City. It is one of the most important and largest brokers of decorative and fine art.

***Spanish Ballet*:** An oil painting created by the French painter Edouard Manet in 1862.

Spies, Werner (b. 1937): German art historian, journalist and organiser of exhibitions. From 1997 to 2000, he was also a director of the Centre Georges Pompidou in Paris. Spies was fooled by art forger Wolfgang Beltracchi.

Stege, Heike: Head of the scientific department for pigment analysis at the Doerner Institute in Munich.

***Still Life with Meadow Flowers and Roses*:** An oil painting created in 1886 by the Dutch painter Vincent van Gogh, currently located in the Kröller-Müller Museum in the Netherlands.

Strontium 90 (^{90}Sr): A radioactive isotope of strontium, with a half-life of 28.8 years, produced by nuclear fission.

Swiss Federal Institute of Technology in Zurich (ETHZ): One of the world's leading universities for technology and the natural sciences.

Switzer, Robert: Dean of undergraduate studies and Director of the Core Curriculum at The American University in Cairo. Philosopher.

Synchrotron–X-ray radiation: X-ray radiation emitted when charged particles are radially accelerated using bending magnets.

Synthetic binders: Formed by mixing polymers, resins and oils.

Tacking margin: Margin formed along the edges of the canvas where the tacks hold it to the wood.

Tanning, Dorothea (1910–2012): American painter, printmaker, sculptor, writer and poet; wife of artist Max Ernst.

Tempera painting: Work executed with pigment ground in a water-miscible medium.

Terpolymer: A polymer formed as a result of the polymerisation of three monomers.

***The Cardsharps*:** Painting by the Italian Baroque artist Michelangelo Merisi da Caravaggio. Caravaggio may have painted more than one version; however, the original is generally agreed to be the work acquired by the Kimbell Art Museum in 1987.

Thiis, Jens (1870–1942): Norwegian art historian, conservator and a prominent museum director. He was conservator at the Nordenfjeldske Kunstindustrimuseum (Museum of Arts and Crafts) in Trondheim beginning in 1895 and director of the National Gallery in Oslo from 1908 to 1941.

Timofeyev, Nikolai (b. 1943): Lieutenant-General, 1st second in command of Commander-in-Chief of Moscow military district of Anti-Aircraft Defence.

Treves, Toby: Former curator of twentieth century British art at Tate, and CEO of Modern Art Press Ltd, a registered charity, which specialises in publishing catalogues raisonnés and other educational books on art.

Turner, Nicholas: Formerly the keeper in the British Museum's Department of Prints and Drawings and Curator of Drawings at the J. Paul Getty Museum. Independent art historian.

Ultraviolet fluorescence: When ultraviolet radiation is absorbed by a substance, it causes excitation of its atoms, which in turn results in the release of visible light.

Ultraviolet radiation: Electromagnetic radiation with shorter wavelength than that of visible light.

van de Wetering, Ernst (b. 1938): Dutch art historian and one of the world's leading experts on Rembrandt and his work. He was the leader of the Rembrandt Research Project for many years and has published the sixth and final edition of the project.

van der Weyden, Rogier (1399/1400–1464): Great Early Flemish artist, together with Jan van Eyck, was one of the most influential northern European artists of the 15th century.

van Dongen, Kees (1877–1968): Dutch painter who was very influential in the Fauvist movement and who was known for his sometimes freakish and strange portraits.

van Gogh-Bonger, Jo (1862–1925): Johanna Gezina van Gogh-Bonger, the wife of the art dealer Theo van Gogh and the sister-in-law of the Dutch painter Vincent van Gogh.

Van Gogh Museum: An art museum founded in 1973 in Amsterdam, Netherlands, that is dedicated to the art works of Vincent van Gogh and his contemporary artists.

van Gogh, Theo (1857–1891): Theodore or Theodorus van Gogh, art dealer and brother of the renowned Dutch painter Vincent van Gogh.

van Gogh, Vincent (1853–1890): World famous post-impressionist Dutch painter.

van Meegeren, Han (1889–1947): Dutch painter, considered to be the boldest and most ingenious forger of the 20th century.

van Tilborgh, Louis (b. 1956): Senior researcher of the Van Gogh Museum.

Vasari, Giorgio (1511–1574): An Italian painter, a writer, architect and historian. He is best known for his important biographies of Italian Renaissance artists. He was born in Arezzo (Tuscany) and died in Florence.

***Vase de Fleurs (Lilas)*:** An oil painting created in 1885 by the French painter Paul Gauguin.

Vasilev, Andrei: Russian art collector.

Velázquez (1599–1660): Diego Rodríguez de Silva y Velázquez was a renowned Spanish Baroque painter and one of the most important painters of the Spanish Golden Age. He excelled in painting portraits which had a very special lifelike quality.

Vellum: Parchment made from calf skin.

Vermeer, Johannes (1632–1675): Famous Dutch painter.

Verrocchio, Andrea del (1435–1488): Born Andrea di Michele di Francesco de' Cioni, was an Italian painter, sculptor, and goldsmith who was a master of an important workshop in Florence.

He was known by his nickname Verrocchio which in Italian means accurate eye as a tribute to his artistic achievement.

Vezzosi, Alessandro (b. 1950): Founder and director of the Museo Ideale Leonardo da Vinci in Vinci, Italy. Italian scholar and artist. An expert on interdisciplinary studies and creative museology and has written extensively on Leonardo da Vinci and the Renaissance.

Viscount Lee of Fareham (1868–1947): Arthur Hamilton Lee was a British politician raised to the peerage in 1918. During World War I, Lee was military secretary to David Lloyd George (1916).

***Vitruvian Man*:** Drawing produced around 1490 by Leonardo da Vinci, in pen and ink on paper. It represents a nude man in two overlapping positions, inscribed in a square and a circle, with arms and legs apart. It is generally referred to as the "Canon of Proportions".

Vyborg Court of St Petersburg: Court belonging to the administrative and municipal district of Vyborg (historical north-western area of St Petersburg named after the castle town Vyborg after the latter was taken from the Swedish Empire by the Russian army under Tsar Peter the Great during the Great Northern War).

Wacker, Otto (1898–1970): A German art dealer who was involved in commissioning and selling forgeries of paintings by Vincent van Gogh. He succeeded in easing the suspicions of leading experts on van Gogh and had them issue certificates of authenticity in spite of an obscure provenance.

Wadum, Jorgen: Director of the Centre for Art Technological Studies and Conservation (CATS) which is a research consortium between the National Museum of Denmark, the National Gallery of Denmark, and the School of Conservation, Copenhagen, Denmark. He is also the keeper of conservation of Statens Museum for Kunst, the National Gallery of Denmark.

Walter, Philippe: Director of the Structural and Molecular Archaeology Laboratory at the University of Pierre and Marie Curie, Paris VI.

Werner, Schade: German art historian and expert on Cranach.

Wheelock Jr, Arthur: Curator of the Northern European Art Collection at the National Gallery of Art in Washington D.C.

***Woman with A Jug*:** Also known as *Portrait of Madame Manet Holding a Ewer*; an oil painting created in 1858 by Edouard Manet.

Wullschlager, Jackie: Chief Art Critic of the *Financial Times*, where she has worked since 1986.

X-ray diffraction: Refers to the scattering of X-rays by a solid. It produces a pattern that gives information about the structure and nature of the crystal.

X-ray fluorescence (XRF): It is a non-destructive technique widely used for the elemental analysis of a particular area in a painting. It makes use of a focused beam of X-rays which impinges on the surface of the painting.

X-ray radiography: A non-destructive testing method that uses X-ray radiation to penetrate the solid object (such as a painting) and produce a radiograph (or image). In case of paintings, the thickness and density of the pigments affect the amount of radiation reaching a detector. Images of underdrawings which are not in charcoal can be observed.

Zinc oxide: White powder used as a white pigment that replaced lead white

Index

www.ingramcontent.com/pod-product-compliance
Lightning Source LLC
LaVergne TN
LVHW020507100826
845148LV00003B/719